It Just Doesn't Add Up

It Just Doesn't Add Up

For more information on dyscalculia please visit www.dyscalculiasupportcentre.com

Published by
Tarquin
Suite 74, 17 Holywell Hill, St Albans
AL1 1DT, UK
Telephone (+44) 01727 833866
Fax (+44)0845 456 6385
info@tarquingroup.com

www.tarquingroup.com
For many ways to make mathematics fun

Paperback ISBN 978-1-911093-00-8
Ebook   ISBN 978-1-911093-01-5

Printed by TJ International Ltd, Padstow, Cornwall, UK

It Just Doesn't Add Up

Explaining Dyscalculia and Overcoming Number Problems for Children and Adults

Paul Moorcraft

For Jenny

Foreword

I met Paul when Marinella Cappelletti and I were carrying out a study of high-functioning dyscalculics. We wanted to find out why they were so bad at arithmetic, while being so good at other things. We also wanted to know how they managed to be successful in this highly numerate society, since there was abundant evidence that people who are really bad with numbers are more likely to have low educational attainment, more likely to be unemployed, more likely to in trouble with the law and more likely to be mentally or physically ill. We tested film directors, TV producers, actors, even a distinguished science journalist. Paul fitted the bill perfectly. Extremely high-functioning. He had, as he quickly pointed out, published more books than me, taught at more universities, and was certainly better-known. On formal testing, however, his numerical skills were a disaster. I couldn't understand how he came to be admitted to university, which required O-level maths when he was a student – with no exceptions. On the basis of our tests, he must surely have failed. Fortunately, in those days the exam curriculum contained more geometry, and it turned out that he was good enough at geometry to scrape a pass. This confirmed my hunch that dyscalculia was crucially about numbers, not about maths more generally. Indeed, I have met world-class mathematicians whom I have reason to believe are dyscalculic. Of course, they don't work in number theory, but in branches more suited to their abilities, such as geometry.

Our high-functioning dyscalculics enabled us to pinpoint the exact cognitive deficits underlying dyscalculia, precisely because they were high-functioning. We could rule out deficits in intelligence (they were all highly intelligent), in memory (they all had good short-term and long-term memories), of education (they were all well-educated), of language abilities (most had written books or presented TV or radio programmes) and ambition (they were all highly successful in their occupations). They also helped us pinpoint the brain network that was abnormal when they were working with numbers.

Paul kindly agreed to take part in our study, which, naturally, meant endless hours doing exactly the kind of things he hated – comparing numbers, remembering numbers, adding and subtracting numbers, and so on.

He also wanted to help with our project of making the disorder better known. This is tremendously important. Around five percent of the population suffer from dyscalculia, and most are seriously handicapped by their disability. Although it's comparable in its prevalence with dyslexia, it is far less widely recognized. As the Government's Chief Scientific Officer, Sir John Beddington, wrote in 2008, 'Dyscalculia is currently the poor relation of dyslexia, with a much lower public profile. But the consequences of dyscalculia are at least as severe as those for dyslexia.' This message has not penetrated the minds of the current government, alas. If you search on the Department for Education website, all you get is a reference to a 2001 document that, you are warned, does not represent government policy.

Up until he met us, Paul had managed to keep his disability a secret – counting on his fingers under the table at board meetings when millions of pounds were at stake – because it could cause difficulty and embarrassment in his working life. But now he was prepared to 'come out' as dyscalculic to help the cause, and he has appeared on radio and TV, and in the press, to show that dyscalculia, though disabling, needn't prevent the sufferer being a success in life.

This book will, I believe, help to raise the profile of dyscalculia, and should be required reading for all sufferers, all parents of sufferers, and all teachers, not to mention ministers and officials in the Department for Education.

Brian Butterworth
Emeritus professor of cognitive neuropsychology in the Institute of Cognitive Neuroscience at University College, London
October 2013

CONTENTS

About the author

Paul Moorcraft has been a life-long sufferer from acute dyscalculia, and has been recorded as having one of the lowest adult scores in numerical ability in the UK. Nevertheless, he has achieved a great deal (though very rarely in maths). He studied at a variety of universities: Swansea, Lancaster, the Hebrew University, Jerusalem, the University of Rhodesia/Zimbabwe and the University of South Africa, obtaining, *inter alia*, a BA (Joint Honours), MA and D Litt et Phil. He was awarded a rare double distinction in his post-graduate education certificate while training to be a teacher at Cardiff University. Although he has taught in schools and worked as an examiner for the Joint Matriculation Board in the UK, he has worked mainly at tertiary level: teaching, full-time, politics, international relations, history and journalism at the universities of Rhodesia/Zimbabwe, Natal, the Witwatersrand, Cape Town, Deakin (Australia), Waikato (New Zealand), Bournemouth, and Cardiff; he also taught part-time at the Open University and Westminster University. He was a visiting professor at Baylor University, Texas, and is a visiting professor at the School of Journalism, Media and Cultural Studies at Cardiff University.

In addition, he has worked for many of the major TV networks as an independent producer, making over 100 broadcast documentaries, most recently for Channel Four News. Besides editing a number of business and security magazines, he was a special correspondent for *Time* magazine in Africa, and writes regular opinion pieces for the quality newspapers in the UK, US and South Africa, as well as being a political pundit for BBC radio and TV, Sky TV and Al-Jazeera. He has worked in over 30 conflict zones as a war correspondent. Also, he has been a senior instructor at the Royal Military Academy, Sandhurst, and the UK's Joint Services Command and Staff College, as well as being an inmate of the Ministry of Defence in Whitehall.

He writes two to three books a year on military history, crime, politics, travel and international relations, as well as being an award-winning novelist. His most recent non-fiction works are: *Inside the Danger Zones* (2010), *The Rhodesian War* (2011)

Mugabe's War Machine (2011), *Shooting the Messenger* (2011) and *The Total Destruction of the Tamil Tigers* (2012). His best-known novel is *Anchoress of Shere*, now in its twelfth edition.

He was also the director of the Centre for Foreign Policy Analysis, London, a think tank dedicated to conflict resolution. He was, for example, head of mission for 50 independent British observers in the 2010 Sudan elections. Professor Moorcraft has also been a paramilitary policeman and an elected local politician, in addition to being a crisis-management consultant to an array of blue-chip companies. The list could go on, but the point is made that dyscalculia can be overcome. Dr Moorcraft has also been involved with dyscalculia issues for nearly ten years, sometimes broadcasting and lecturing on the subject, as well as talking about the disability in schools. He is not only profoundly number-blind but, incidentally, now almost completely blind. So this book is based on actual experience of achievement based on overcoming very real disabilities.

Why did I write this book?

I have suffered from acute dyscalculia all my life, though I didn't have a name for the condition until fairly recently. I thought I was just lousy at maths. It has affected my whole life. Nevertheless, I went from school to teaching at ten universities around the world; from primary to professorship. So the message of this book is entirely positive: just because you can't count doesn't mean you don't count.

In a BBC programme, *A Child of our Time*, with Professor Lord Winston, I was described as a 'super-achiever'. That was kind, but certainly an overstatement. I have achieved a great deal in a number of fields; though I was always bad at making – or for that matter, counting – money.

This book describes what dyscalculia is, what it does, and how to cope and then overcome the disability, whether in a child or adult. It is not, I hope, too technical; rather it is intended to be a practical guide for the general reader.

Dyscalculia is a real disability. It is not some fancy fad, an excuse for performing badly. It exists. Unlike its near-neighbour, dyslexia, comparatively little is known about the condition. This book will talk about some medical and scientific issues, but – to repeat – it is a can-do type of book, not a scientific treatise. And I have avoided using maths formulae almost completely. Honestly.

I knew I had a problem in primary school: I had trouble with my times table, and doing simple mental arithmetic. It mattered less then, because I was considered very bright. Despite my relatively poor working-class background, I managed to pass the eleven-plus exam to go to a good grammar school in Cardiff, Wales, where I excelled at most subjects – but not maths.

My maths teacher – 'Hank' was his nickname – despaired of my ever passing the GCE exams when I was 16. Without maths O-level I could not go on to university, which I thought was my best way out – if not from the ghetto – then at least from my so-called 'council estate'. Hank used to call me lazy or occasionally stupid, but no matter how hard I tried I couldn't cope. And, I must admit, by about 14 or 15 I had given up – I couldn't understand the principles of algebra at all. But somehow I knew I had to pass my maths exams.

Extra coaching was not an option in my single-parent home – I lived with my occasionally employed divorced mother. Then dyscalculia stepped in to help. One of the aspects of acute dyscalculia is often difficulty in planning and managing time. I was frequently late for school so spent much time in detention for unpunctuality, as well as sometimes for bad behaviour. Hank used to get me to write out the geometry theorems in detention and for other failures in my maths classes. He believed that geometry keeps you in shape. The extra burden was heavy, and Hank had a reputation as a sarcastic bully. Yet I sometimes wonder now whether he was trying to help me.

I scraped just over 50 per cent average, a pass at O-level maths. I understand that I managed to get around 90 per cent in geometry, which was just enough to boost my appallingly low marks in algebra and arithmetic. Geometry, being more about shapes rather than numbers, made more sense to me. And I had a very good memory, so I could recall the theorems, not least after endlessly writing them out in detention. I don't know whether my good memory was innate – it is not so good now – or whether it was one of the coping mechanisms my mind developed to help me survive with dyscalculia; though I couldn't remember numbers at all. One exception was, and is, the registration number of my first motorbike.

I went into the sixth form and savoured the joy of dropping science and maths, to concentrate on the arts subjects such as English and history, subjects that I enjoyed and was good at. Because my mother remarried, I left home while I was still at school and moved in with an aunt. Life was not easy, but eventually I secured sufficient A-levels to study at Swansea university in 1967.

I was a good student and learned to schedule my time better. Like most dyscalculics, however, I was often hopeless at estimating how long tasks would take. I soon learned to make lists of things to do, and work out a study and personal timetable. I even found time to do voluntary work (as well as working for money in all my vacations and driving a taxi part-time in term time). I used to help young adults to overcome their illiteracy. What struck me then was how much effort they put into evasion tactics, understandably refusing to admit their deficiency to peers. I used to tell my students: put half that effort into trying to learn to read and write and you will succeed. At that time I didn't know anything about dyslexia – a form of word blindness – and simply put it down to bad schooling in the deprived areas of South Wales, lack of motivation at home, or – to be frank – that some were just lacking the intellectual capacity.

I successfully avoided anything to do with numbers at college, though I couldn't avoid the social ramifications of my disability. I grew up without a phone in the house, and we certainly didn't have one in any of my student flats. But sometimes I would meet a young lady with a phone. At the end of a first date – if all went well – I would somehow have to ask for her number, and she would sometimes tell me. I knew that it would be gone from my mind in seconds. I used to say something along the lines of:

'Sue, you've got an unusual name, do you mind writing your full name and number down for me?

'What's so unusual about Sue Jones then?'

'Never mind, just write it down for me. I don't want to forget it,' I would say with a forced smile.

This was one of my teenage coping tactics for getting around my number problem.

I graduated from Swansea, and then pursued an MA at Lancaster University, where I worked in a pub in the evenings. We didn't have calculators or fancy tills. I had to work out the change for myself. I was not very successful at this, but I tried to make up for it by my amiability with customers and sheer hard work.

Device for dyscalculics experiencing spacial disorientation.

I returned to Cardiff, to study to become a teacher at my local university. I continued driving a taxi in the Welsh capital, and that indeed was my first full-time job after graduating from three universities. I knew the city well, and was good at navigating my way around. I was lucky in that I did not suffer from spatial disorientation, which I learned much later was common in dyscalculics. I had a good sense of direction. Later, at the Royal Military Academy, Sandhurst, I soon discovered that I was less good at orientation with maps, perhaps because of the numerical bearings.

I studied at other universities and obtained a doctorate. I also taught at universities around the world in a variety of subjects, though the accolade that pleased, and surprised, me most was becoming a visiting professor at my home university, Cardiff. This is not an autobiography. I just want to make it abundantly clear that dyscalculics can overcome their disability, providing they adopt the correct strategies. I will use further examples from my own experience and that of other dyscalculics throughout the book. Here I want to mention just a few obvious symptoms.

Working as a foreign correspondent, I often had to use the phone, of course, and to be able to take down phone numbers, often quickly and sometimes in foreign languages. I would transpose numbers – jumble up the order – in many cases. Because of two years at Sandhurst, and other military experience, I used to pretend to use a quasi-military technique of asking for numbers. I would rarely get the number first time:

'Would you confirm that number?

'537 – confirm 537.'

I didn't often say 'Roger' or other military communications protocol, but my style was akin to listening on a bad field telephone. I would use a staccato brief military approach and break up the numbers into threes.

It worked most of the time.

This did not, however, transfer to the need to remember codes for safes and locked doors in the Ministry of Defence, when I worked in Whitehall. For security reasons you were not allowed to write down key numbers. I coped by delegating someone else to lock and unlock the safe, but made a friendly fuss about supervising this ritual at the end of the day. Sometimes I would cheat and write down codes. And of course I had to do the same with PIN numbers, not least for my bank cards.

As you can imagine, doing accounts was almost impossible. In 1984 I organised three trips behind Russian lines during the war in Afghanistan. In those days, Britain and America were on the same side as Osama bin Laden. I was the field producer of a documentary on the war, and led (badly) one of the most difficult penetrations into the fighting zone. Inside Afghanistan, I had to organise, for example, payment for the horses we used. That involved Afghan currency. Beforehand, we had to kit out the film crews in local Pakistani dress, another of my responsibilities and another currency. We also filmed in the UK. Some of the payments were in US dollars. I had to reconcile the whole lot in Johannesburg, in South African rand. Nobody could get paid – for three months' hard and dangerous work – until I signed off the accounts. I had a calculator, but it took me ten times longer than a person with normal maths skills. Despite my care and attention, I made a mistake with one, crucial, digit on the final sum. These weren't the exact figures, because I can't remember, but instead of writing, say, £250,000, I added an extra zero. And that made a big difference. Luckily, the film company's accountant knew me to be an honest fellow and assumed I had made a typographical error, though the mistake delayed, for a few days, the final payment for me and my war-weary colleagues.

When I set up small businesses in the UK, I employed an old college friend as my accountant. Doing my personal and business accounts became an annual nightmare. One way of half-coping with this issue was to work on my expenses and accounts nearly every day, writing them by hand in an old-fashioned ledger. Doing a bit every day, keeping up to date, did help me and my accountant.

About ten years ago, my problems with maths came to a head. I had hidden my deficiency my whole life. You can't admit to girlfriends, future employers or business colleagues that you can't count. They might laugh off problems with phone numbers, timetables, PIN numbers etc, but when it came to the bottom line in business, well it was always the bottom line that mattered.

A friend from Durban, South Africa, had left a message on my answering machine, urgently asking me to respond. It was Sod's Law, but my machine was playing up and would wipe messages after playing them back ONCE. As usual, I was poised with a pen and paper, but I couldn't get the whole long international series of digits. I had swapped around one or more numbers, as irate and colourful responses from sleepy strangers in South Africa soon confirmed. When my South African friend returned to London a week later, she angrily remonstrated with me about failing to respond to her message. After years of hiding my problem, I decided to explain, for the first time.

She didn't believe me. 'Typical of your bullshit,' she said. 'I bet you can't even come up with a name for your so-called disability.'

I couldn't; so next day I rang a very old friend, Liz Ainley, who specialised in teaching dyslexic and dyscalculic children. I confided in her and she was most helpful. She suggested I email an expert in the field, Professor Brian Butterworth. Liz told me to outline briefly my academic background, to contrast with my acute mathematical weaknesses.

Brian responded promptly and kindly, and invited me to undergo a series of tests at the Institute of Cognitive Neuroscience in central London. Although the tests were designed for young children, I scored very badly – even though I tried my best, and I can be very competitive in such things.

'You have achieved perhaps the lowest score of any adult I have tested,' Brian said, baldly. 'Frankly, you appear unemployable, even as an agricultural labourer.'

He knew of my background, my too-obvious self-confidence, and we had established a rapport, so his tough comments must be put in context.

I replied: 'Well, Professor, I'm also a professor, and I have written far more books than you have.' My competitiveness again.

'That's precisely why I want you to help us with our research and to generate more public awareness of the condition.'

Brian asked me if I would go on TV to discuss my problem.

I said I needed to consult with my business colleagues. I had just persuaded a company to invest a relatively large sum of money in a publishing venture, involving security magazines. We had produced a detailed business plan, and I had been the one who had initiated the deal. It was not a good time to admit that I was a financial incompetent. Nevertheless, my colleagues agreed when I said my main purpose was to help children with the condition.

Brian sent me an email, using the language I had deployed when I told him I wanted to ask my colleagues. The email was headed: 'Coming out.'

One of my colleagues was gay, and a somewhat crusading gay to boot. His desk was adjacent to mine. He saw the email heading and declared to the whole office: 'If you're gay, Paul, I am giving it up now – you've just made it deeply unfashionable.'

I did come out on the major BBC programme, *A Child of our Time*, anchored by Professor Lord Winston, and did appear on other programmes as well, with Brian. Friends joked that it was all about setting up a pre-emptive defence in case of a tax investigation. It wasn't, of course. Despite energy, and perhaps flair, as well as damned hard work, I never made enough money to prompt any kind of tax avoidance.

One frequent question, however, did trouble me: 'If you were labelled a "dyscalculic" as a child, would that have prompted you to just give up,and admit defeat?'

It was certainly a valid question that troubled me for a while. When I was in primary school in the early 1950s, no teacher, at least in Wales, would have understood the term. But if my mathematical weaknesses had been formally diagnosed, would I have accepted that I was somehow 'challenged', and perhaps stayed on in one of the many manual part-time jobs I did in the school holidays or part time in the evenings? My first job was a grave-digger-cum-graveyard-maintenance man.

I can't answer that with absolute certainty, though nobody is better qualified than I to make a stab, at least about myself. I genuinely wish I had understood what was wrong with me much earlier: to know that I wasn't stupid. If I'd had some kind of road map of the rough terrain, the routes around massive potholes and dangerous mountain ridges, it would have saved me a lot of heartache. I would have avoided many falls and wrong turnings.

I want to help dyscalculics but, although I have included many practical suggestions, this is not a classic self-help book. Most of these types of books are written as though the reader is being asked to join a church or political party. I am not trying to convert anyone, but I do want dyscalculic adults and children to find a way to overcome a condition I have battled with my whole life.

The more you know, the more you can cope. There is no one simple silver-bullet 'cure' for dyscalculia, but many good methods of alleviating the symptoms do exist. If just one child can learn from my many mistakes, this book will be worth it for me.

A number of people helped me with my journey into the heart of dyscalculia. If I may name a few. Liz Ainley helped with editing and introduced me to Professor Brian Butterworth, Britain's leading authority in the field. He has always been most helpful, including kindly agreeing to write the foreword. I also met Dr Marinella Cappelletti at London's Institute of Cognitive Neuroscience; she was a post-doctoral fellow in Brian's lab. I was her guinea pig in a number of her research tests. It was our discussions that inspired this book. I thank her for her advice on the more scientific aspects of dyscalculia. I must also express my gratitude to Jennifer Muddiman, an experienced maths teacher from whom I learned so much about trying to get children to handle maths. She guided me through the dark intricacies of educational bureaucracy, as well as literally guiding me around many other things whenever I deigned not to use my white cane. The cartoons were magically conjured up by Dave Frewin. The book designer was my ever-patient friend, Tony Denton. Finally, I thank my agent, Susan Mears.

Now that it is finished and ready for publication, all I need to do is to check that the chapter headings and pages are in the correct numerical order!

Professor Paul Moorcraft
Surrey Hills, England

September 2013.

1: What is dyscalculia?

D yscalculics have difficulties with numbers. It is a learning difficulty in mathematics. Sometimes it is termed 'mathematics disorder' or 'mathematical disabilities'. As we shall see later, dyscalculia is also associated with other difficulties besides purely handling numbers, such as time management and spatial reasoning.

Dyscalculia is often associated with dyslexia – difficulties characterised by problems in reading, writing and spelling. Dyslexia was identified much earlier than dyscalculia and is much better understood. When diagnosed, especially in early childhood, targeted teaching can help a great deal. If not, frustration and low self-esteem can set in, and literacy problems can continue throughout life with obvious impacts on employment prospects.

Often, but not always, children can suffer from dyscalculia *and* dyslexia, but this book is concerned primarily with problems with numbers. I shall draw sometimes on the more advanced work done on dyslexia, but it is a separate issue – though in some people they may be related. Generally speaking, research on dyslexia is about 20 years ahead of that done on dyscalculia, so we still have much to learn, and many opportunities to discover how to cope better with the disability. Studies have suggested that suffering from dyscalculia makes it more difficult than having dyslexia to be successful in the world of work.

It has been estimated that low numeracy costs the UK £2.4 billion per year, mostly in poor productivity. At least 25 per cent of adults have maths deficiencies in the general UK population.

The word 'dyscalculia' comes from Greek and Latin and means 'counting badly'. The prefix 'dys' comes from the Greek and 'calculia' derives from the Latin 'calculare' – to count. 'Calculare' relates to 'calculus', which means a pebble or one of the counters on an abacus. So now you know.

The term 'dyscalculia' is a popularised, now generally accepted, form of describing a maths disability. It has been in general use since the 1960s, at least in the US. The American National Center for Learning Disabilities has defined dyscalculia as 'a wide range of lifelong learning disabilities involving math'. Researchers as far back as the early 1800s used terms such as 'arithmetic deficit', and so on. Today the media sometimes refers to 'math dyslexia', 'number blindness' or 'digit dyslexia'. But I shall use dyscalculia, as the most readily understood and handy term for the problems – and solutions – discussed in this book.

Dyscalculia is real, not some American fad. I've had it my whole life. Teachers know it is real. And it is recognised by major medical organisations, such as the World Health Organization.

How common is dyscalculia?
Estimates vary, according to the definition and nature of the research. In the UK and the US, the figure is usually given as 5-7 per cent of the population. (Although some estimates go as high as 11 per cent, Brian Butterworth's research in Cuba suggested a low of 3.6 per cent.)[1] So, in an average class of, say, 30 children, one or two are likely to suffer from the disability. In Britain, then, more than 3 million people are likely to be dyscalculic, but the vast majority have not been diagnosed as such.[2]

Many people suffer from dyscalculia, but they can still achieve great success. That is the main message of this book: to understand, to cope and to achieve. My two favourite programmes on TV in the 1960s were The Mary Tyler Moore Show and Happy Days. Mary Tyler Moore became a big TV star, despite her dyscalculia, as did 'The Fonz', Henry Winkler. Henry had a double whammy – dyscalculia and dyslexia, as did pop singer Cher. A more recent example is Mick Hucknall, the lead singer of the UK band Simply Red. He sold millions of records and made millions of pounds, but struggled with dyscalculia. We are told that even Albert Einstein had problems with basic maths, although that may be apocryphal. Benjamin Franklin and Thomas Edison both left school at 12 because of their learning difficulties, not least in maths.

It is often said that dyslexia can encourage some sufferers to super-achieve – the list is endless: from President John F Kennedy, to John Lennon, Virgin Boss Sir Richard Branson to film producer

Steven Spielberg. Though there is less research-driven evidence, I am convinced that dyscalculia can also be overcome and, given enough motivation and energy, spectacular success can sometimes be achieved.

Although more males show evidence of dyslexia (70 per cent compared with 30 per cent of females), with dyscalculia, studies suggest the ratio is 50:50. Despite this evidence, Jenny Muddiman, an experienced teacher originally from Hampshire, England, told me that gender plays an important role in dealing with maths. She said: 'It's almost a feminine thing to say, "I'm not good at numbers." Mothers say: "I tell my kids to ask their dad." Maths is perceived as a more of a male thing.'

She thought that the media perpetuated this. 'They say girls are catching up, or outperforming boys in maths and sciences. The media are making it a gender issue.'

As we shall discover later, issues such as map-reading, or even indicating left and right, may often be associated with gender biases.

What causes dyscalculia?

Like dyslexia, dyscalculia is widely recognised as having a genetic component. For example, if one pair of twins is dyscalculic, the other twin is more likely to be dyscalculic if the twins are identical rather than if they are non-identical. Parents or siblings of dyscalculic children are ten times more likely to have the condition, compared with the general public.

Scientists now generally assume that maths ability resides primarily in the parietal lobes of the brain. It seems that these systems are abnormal in dyscalculics, although work on the actual physical cause and the hereditary factors has not yet produced one definitive medical explanation. And so dyscalculia has no definite 'cure' either.

Some researchers claim there is a correlation with premature birth. Some adolescent dyscalculics who had been born prematurely have lower gray matter density in the left intraparietal sulcus, the part of the brain where activity takes place during mental arithmetic. Other research has indicated a possible correlation with abnormality in the X chromosome.

Brian Butterworth believes that it is at heart a deficiency in basic number sense and not of memory, attention or language, as other experts have suggested. He says that learning basic arithmetical facts can seem like learning to repeat sounds in an unknown language for dyscalculics. 'We are beginning to understand how the brains of dyscalculic learners are different from typical learners, ' he said, 'but we still do not know why they are different.' And it is not yet clear whether remedial intervention makes the dyscalculic brain more normal, or whether intervention helps the dyscalculic learner find a different way of doing the same mathematical task.

Despite the academic debate and the fact that some primitive cultures have little or no sense of numbers of more than four or five, it is probable that we are born with an innate sense of numbers, but in dyscalculics that is impaired. We will be concerned with dyscalculia as a 'developmental' difficulty that a child may be born with. I will not be dealing with so-called 'acquired dyscalculia' that may have resulted from external injury or stroke. 'Acalculia' is related to serious brain damage, where there is no understanding of numbers, even counting to ten. But there are milder and more specific cases: for example, one man lost the memory of his times tables, but had no difficulty with addition and subtraction. I will be discussing only developmental dyscalculia (abbreviated to DD in scientific literature), and how it is possible to learn about how numbers work, to help children and adults deal with poor maths.

Dyscalculia occurs in people across the entire IQ range. Often, they do very well in other subjects, besides those involving maths. Even with maths – and without getting too technical – dyscalculics may suffer from arithmetic difficulties (such as calculation and number fact memory), do not suffer, or may even be very gifted, in abstract mathematical reasoning (though this is hard to test, and prove). Possibly, this was the case with Albert Einstein, but not – unfortunately – with me.

KEY POINTS
♦ 'Dyscalculia' is the most convenient name for problems with maths that relate to natural development and not accidental brain injury
♦ Probably largely genetic, it affects the parietal lobes of the brain and reduces number sense
♦ It is a real disability recognised by the World Health Organization It occurs in approximately 5 per cent of the UK population. Each school class is likely to contain at least one dyscalculic
♦ It affects males and females equally
♦ Sufferers can achieve a great deal, and it is not related to general IQ levels.

2: What are the key symptoms?

Strewth! Well at least I can count the symptoms I haven't got. ZERO!

Let's look first at obvious symptoms, which are likely to be exhibited early on in school. The manifestations of dyscalculia are many, and they are sometimes confused with other conditions such as attention-deficit disorders. The condition, once properly diagnosed, can vary from mild to severe. Although it is generally a lifelong difficulty, many coping mechanisms exist.

Maths problems at school
♦ Taking an bnormally long time to tell which number is larger or smaller
♦ Trouble rounding up even fairly low numbers
♦ Reliance on strategies such as finger counting
♦ Confusion about mathematical symbols: +,-, ÷ and x
♦ Difficulty with addition, subtraction etc
♦ Sometimes transposing/reversing numbers – writing 63 for 36, for example

♦ Sometimes confusing similar numbers, for example 3 and 8
♦ Poor mental arithmetic skills
♦ Trouble conceptualising basic formulae
♦ Problems with remembering maths operations – mastering it one day and completely forgetting it the next
♦ Inability to remember times tables
♦ Problem using calculators. Dyscalculics have to check a number of times – until the answers are the same two or three times
♦ Inability to remember numbers, even the phone number at home
♦ May have difficulties in reading a clock
♦ Sometimes difficulty in copying shapes accurately
♦ An inability to guess quantities (without counting) even with small quantities
♦ Difficulty in counting backwards from ten
♦ Many children with dyscalculia are not good at puzzles, which may suggest they have visual-spatial issues
♦ Difficulty keeping score during games. Football is easier, but cricket and tennis are more complex. They may have a good grasp of the principles of the games, but find it hard to plan ahead more than a few moves, in chess, for example.

The above are just a few of the issues dyscalculics will struggle with at school. When they become older, they sometimes overcome, or learn to evade, many of the difficulties. More often – if the problem is not addressed when young – they will resort to disguising their deficiencies. Adults, for example, will rarely count on their fingers in public (though I do sometimes, when I forget that people may be watching me!).

For a full list of symptoms for children and adults, see '50 + symptoms of dyscalculia' in the appendices.

Time and space problems
♦ Difficulty with telling the time. Conceptualising time and the passage of time when numbers are involved
♦ Dyscalculics will often be chronically unpunctual, especially when they are younger – before work disciplines often enforce a new rigour

♦ Judging distance
♦ They may have issues with comprehending or picturing mechanical processes
♦ They may struggle to visualise geographical locations – from states/counties in their own country to lay-out of streets in their own neighbourhood. In short, they often get lost, even in school. Although this behaviour is not typical, it can be disconcerting.

They may have problems with grasping north, south, east and west, even with a compass or map. Most young children have problems defining left and right, but it is much worse for dyscalculics. On into adulthood, many will still have a difficulty with defining left or right, especially under pressure, for example when giving directions to a driver. Dyscalculics will tend to have difficulties using a map.

In summary, a nine-year-old dyscalculic will have on average the same maths level as the typical six-year old. Generally, dyscalculics will plateau at the end of primary school and, without proper intervention, may make only one year of progress in secondary school.

Maths phobia
Sometimes young dyscalculics will develop a phobia about maths and just give up. The secret here is for the parent or teacher never to give in to the desire to label the child dumb or lazy. The children have usually done their very best for years, and just simply can't – with all the effort in the world – get to grips with numbers.

As one Australian teacher said of her own child, Lucy: 'Her writing was always very good; it's the maths that has been the issue. We sometimes have to struggle at school with making people understand that it's not because Lucy doesn't want to do it or can't be bothered. It's because she really cannot do it. She's unable to retain a lot of information.'

In my own case, I felt as though I was being asked to speak in an unknown foreign language for every minute of my arithmetic and algebra classes in school.

Some sufferers, young and old, develop outright phobias about numbers and doing any kind of maths. A note of caution is required here. Although many dyscalculics develop a phobia of maths, you can have a maths phobia without suffering from dyscalculia. Those in the latter group may add to the earlier – inaccurate – assumption that dyscalculia is simply a made-up mind-set, a psychosomatic condition that can be overcome by simply getting rid of the fear. In genuine dyscalculics, such an approach would be like saying a dyslexic had only to overcome his or her fear of spelling. Real phobia based on dyscalculia may induce genuine or feigned sickness and then truancy, all related to maths-avoidance behaviour.

Symptoms of maths avoidance behaviour.

Later on in the book I will look at ways of testing for actual dyscalculia, to screen out phobias in those who do not suffer from genuine numerical disorders, that is those that are genetic and developmental rather than psychological.

Case study: Overcoming dyscalculia

Samantha Abeel was a young American who was considered very smart, even gifted, at school, but she hid her weaknesses: she couldn't tell the time, remember the combination of her school locker or count change in a shop. And she was hopeless in maths classes. She started having anxiety attacks, that manifested as stomach aches, and losing sleep, and distancing herself from her friends. When she was 13, she found the courage to confront her problems and was diagnosed with dyscalculia. She realised that her guilty secret was not her fault.

Samantha was always good at creative writing, despite some problems with spelling and grammar. That was her saving grace in school. She wrote a very moving account of her struggles with dyscalculia in her memoir, *My Thirteenth Winter* (see further reading list). She describes how her mind went blank when confronted with numbers. 'I fight to compute the numbers, but my brain feels as if it is searching through empty file cabinets.' When the teacher wrote numbers on the board, she was lost.

I feel as if I am staring into the face of someone I should know but can't seem to remember. No matter how hauntingly familiar the figures are, they continue to remain anonymous strangers to me and a wave of guilt and embarrassment moves through me.

Samantha always felt alone, even though she had supportive parents and attended a good school. She writes about how she felt denial, anger, isolation, fear, guilt and shame until diagnosed with a learning difficulty. Then she felt relieved and 'free to accept my learning disability as part of my life'. Thereafter she came to view her dyscalculia as 'a rather strange and unusual gift'. She developed all sorts of coping and survival tactics that allowed her to transcend her disability and to reach university. She says that her disability afforded her a 'unique view of the world'. It forced her to always look for, and be open to, creative solutions and not to take the simplest of tasks for granted.

Adult dyscalculics
Testing for adult dyscalculia

Adults will probably not undergo any of the diagnostic screeners used in schools, so they may find the following check list useful. Tick the ones for which you answer 'yes'.

❏ Do you habitually offer bank notes in a shop rather than work out change for a small purchase?
❏ Do you have problems reading maps (especially grid references)?
❏ Do you avoid reconciling your accounts and personal finances at the end of every month?
❏ Do you use just one PIN number for all your cards?
❏ Does it take you a long time to remember your own mobile phone number?
❏ Do you habitually forget passwords?
❏ Do you have difficulties with working out north/south/east/west or even left and right?
❏ Do you always delegate somebody else to work out bills in a restaurant?
❏ Do you often forget people's names, even when you know them quite well?
❏ Do you have to rely in detail on diaries/ personal organisers etc to remember appointments?
❏ Do you often lose important items such as keys?
❏ Do you have a phobia about maths, especially mental arithmetic?
❏ Are other members of your family very poor at maths?
❏ Do you have difficulty in recalling maths facts from school?
❏ Do you avoid checking your bills, especially credit card bills every month?
❏ Do you have problems working out recipes using detailed weights and measures?
❏ Do you have difficulties in working out DIY details – how long a piece a wood or how large a pane of glass must be?
❏ Do you wince at working out VAT on a tradesman's estimate?
❏ Do you struggle with working out the odds and what you might win on bets for the Derby or who will win a political leadership contest?

What are the key symptoms?

❏ Can you easily convert your weight from imperial to metric (stones to kilos)?
❏ Do you avoid internet price comparison sites?
❏ Do you habitually fail to work our your weekly/monthly budget?
❏ Do you find you don't fully understand how mortgages work, especially the advantages/disadvantages of tracker mortgages?
❏ Do you tend to avoid using bus/train timetables?
❏ Do you have problems working out exchange rates for foreign currencies?
❏ Do you often hesitate when working out the tip in a restaurant?
❏ Do you avoid checking your supermarket shopping receipts?
❏ Are you slow in working out sales discounts in shops?
❏ Do you have difficulty remembering when you are told what your blood pressure or cholesterol level is?
❏ Do you let someone else work out the cost of your planned holiday?

This is not a scientific survey – individual failures may in some cases relate to memory problems or simply being a busy parent! But it is a valid indicator of symptoms of dyscalculia if a number of the problems congregate together. If you answer yes to more than 10 of the questions, you are likely to be poor in maths. If you score more than 20, then you are probably dyscalculic. If you agree with all or nearly all 30, then you should think seriously of taking up adult education, not least in maths. Or sitting in with your children when the coach you have paid for pitches up for their extra lessons! Better still, read this book very carefully.

As the dyscalculic grows older, they will exhibit an inability to grasp financial planning or budgeting, sometimes even at a basic level such as checking change in a shop or balancing a chequebook. They will typically think short-term when it comes to financial planning – they will often fail to grasp the big picture.

They will have a constant tug of war about writing down their PIN numbers versus observing good security and trying to memorise them. Typically, they will use only one code for all cards, and then sometimes get anxious about forgetting it, especially if they try to memorise two numbers for two cards.

As adults, they will often feel awkward in restaurants when it comes to working out the bill, and the right percentage to tip. They will nearly always delegate somebody else to do this. They are often self-conscious when checking change in shops.

Dyscalculics often have trouble with remembering names (they would have already given up on remembering phone numbers). They will sometimes get the right starting letter, but say Ronnie instead of Reggie, for example.

Dyscalculics tend to lose things, especially keys, and are often considered absent minded. That was one incentive for me to become an academic – I had the classic excuse of being an absent-minded professor.

To cope, dyscalculics often become very reliant on diaries, and lists of 'things to do'.

Dyscalculics may sometimes have an over-sensitivity to noise, smell and light. They might have an inability to tune out unwanted information or impressions. Typically, they will read in silence, not wanting to be distracted by a radio or TV.[3]

There may be some truth in the old adage that musical and mathematical ability are aligned. Dyscalculics may struggle with formal musical education, especially sight-reading music, though they may be gifted enough to play by ear.[4]

I won competitions at dancing, but I could rarely follow changing physical directions in dance classes. Remembering dance sequences was almost impossible for me. But I could dance (well), and play some musical instruments (badly), more by instinctive feel, rather the ability to understand the set rules or planned choreography.

Here I might even venture that dyscalculics who get on in life tend to be mavericks, because they have to find ways around set rules.

Well, that's my excuse for being considered a troublemaker for most of my life.

Case study: Catriona Fletcher

Catriona Fletcher, a dyscalculic who has been very successful in the fashion design business, agreed with the idea of a maverick tendency. 'Yes, I recognise that – it rings a bell, I do see things differently. It sometimes helps me to shine because I have different views. I enjoy that, and it is useful in my work. It has helped me develop a more independent, creative mind.'

She added: 'Since I have discovered dyscalculia and other people who suffer from it, I realise I am not alone.'

Catriona also explained that she had synaesthesia, an ultra-sensitivity to colours. 'To me colours are numbers, that's how I remember phone numbers.'

The elegant Scot had initially worked as a designer where her sensitivity to colour proved a great strength.

Synaesthesia is relatively rare. Some see numbers or letters as colours; one magazine journalist explained that she saw days of the week and months as misted complex hues. Occasionally taste, smell and music produce colours. Vladimir Nabokov, the Russian author, wrote lucidly about his ability to see letters as colours. But most synaesthetes have great difficulty in explaining their inner rainbow.

Catriona experiences what is called a grapheme variant of the condition, meaning that letters or numbers are inherently and involuntarily coloured. Typically 'A' tends to be red, while 'O', for example, is white or black. Most synaesthetes say the experience is neutral or pleasant; some artists see it as a distinct advantage. The condition has been examined since the early nineteenth century, but research faltered in the mid-twentieth. It seemed to

---

come back into vogue in the 1960s, although interest in the LSD drug may have coloured, literally, scientific enquiry. More recently, it has been estimated that 60 variants of the condition exist. It has been argued, but it is scientifically unproven, that one in 23 people in the UK suffer from (or enjoy) the condition across a range of variants and extremes. A recent scientific survey conducted by the School of Behavioural Science at the University of Melbourne in Australia suggested that the condition is much rarer: in adults, 1 in 1,150 females and 1 in 7,150 males. This survey also indicated that synaesthetes are more likely to be involved in artistic pursuits.

Positives not negatives

The lists of dyscalculic symptoms discussed earlier may depress sufferers and their families and teachers. It is important to understand what the condition means – that is the first step to overcoming it. Very few dyscalculics will have to endure ALL the symptoms – that would be very rare indeed.

Many of the issues – take unpunctuality – can be addressed by determined self-discipline. In my case, my earlier broadcasting career soon taught me that even two or three seconds of 'dead air' on live radio (let alone live TV) can be disastrous. You may do it once and, if you are not sacked by your TV/radio station, you will never do it again. (Though I did once repeat the error, but I had a damned good excuse ready – I was reporting in a war zone, and was under fire.)

Also, most dyscalculics, at least in their adult life, are able to perform some operations with numbers. So they can decide which of two items is cheaper, set the table for the correct number of guests and perform some basic arithmetical operations. What is most common, though, is that these activities usually take a very long time to perform, much longer than people without number problems. This may not seem a big deal, but there are plenty of situations where being abnormally slow becomes an issue. Checking the change in a shop while people queue behind you, sharing the bill at the restaurant while people wait to know how much to pay, checking if you have enough cash to pay or whether you should use a card, remembering a PIN at the cash machine, and so on. In some of these, and many other cases, being slow may be embarrassing or even dangerous (it is not safe to wait minutes at the cash point while trying to remember a PIN).

The good news, however, is that despite the long time required to do them, some of these numerical tasks can be performed. It simply means that the way they are performed is more convoluted or may consist of more basic strategies – for instance counting on fingers – that slow you down. It is essential, though, to remember that these operations are possible to some extent and that it is speed that needs to be improved.

As with any disability, often the brain compensates. Dyscalculics will always have some trouble with maths, but they may be able to get around it, just as colour-blind people learn to manage. Completely blind people sometimes develop extraordinary audio awareness. They can recognise friends by the sound of their walk, for example. Dyslexics often become super-achievers in business, arts and entertainment. Sometimes finding a way around the challenges will become second nature. I ran a number of businesses and raised money for various business plans. I needed help on the financial details, but usually coped with bigger strategic challenges. So 'fixing' these issues can be done by the individual through effort and willpower.

In other cases your own brain will do it automatically. There is less evidence of this with dyscalculics, perhaps because so little research has been done, until very recently.[5] However, the good news is that there is some evidence of accelerated language skills – verbal, reading and writing. Poetic and artistic ability may also be enhanced. Dyscalculics may have excellent memory skills for the printed word. In my own case, I could flick the pages in my mind when it came to Latin translations in English.

Dyscalculics may even be good at science – until a level requiring higher maths is reached. They can be good at geometry. Because it deals with shapes and logic, I managed to secure very high marks, though any algebraic formulae would leave me cold.

Often dyscalculics will have a well-developed sense of imagination, possibly as a cognitive compensation for their mathematical deficits. But I promise you I didn't imagine my high marks in geometry.

Often dyscalculics will have a well-developed sense of imagination.

KEY POINTS
♦ Maths deficiencies at school – from counting on fingers, reversing numbers, failure to learn times tables – may be symptoms of dyscalculia
♦ Poor sense of direction and spatial awareness may also indicate the disability
♦ Very few dyscalculics will display all the classic symptoms
♦ Adults typically have problems with finance, from counting change in a shop to remembering PIN numbers
♦ Successful dyscalculics tend to be mavericks and find imaginative ways of coping.

3: Related or similar disabilities

A cute dyscalculia is not usually diagnosed on its own. Approximately 40 per cent of dyscalculia sufferers are also likely to have attention deficit disorders (ADD) or with hyperactivity (ADHD). They are separate conditions but they are also frequently associated, or 'co-morbid', to engage the scientific/ medical jargon. Again, though they are different disabilities, dyslexia is often found with dyscalculia. Other problems such as auditory processing deficits may accompany the condition. Dyslexia, and probably dyscalculia, are often undiagnosed in deaf children. Like dyspraxia (which affects motor skills) and dyslexia, dyscalculia tends to be genetic. I repeat: pure dyscalculia – suffering from this one disability only is relatively rare. Co-morbidity, also called concurrence, may be as high as 70 cent. I was lucky, in one sense, in that I suffer from pure dyscalculia. Otherwise, I would not have opted for writing as a major part of my career.

Dyspraxia

Also known as DCD (Developmental Coordination Disorder), dyspraxia mainly affects motor coordination skills. The pupil is likely to be very clumsy and poor at planning and organising. They probably won't be good at sports that require balance. In maths, dyspraxic students will struggle to use protractors or set of compasses. Their work will be messy and difficult to read.

Most dyscalculics will tend to count with their fingers, but dyspraxics may avoid doing so, even if they are poor at maths. They may not actually be aware of how many fingers they have and where they are placed (referred to as 'finger agnosia'). This, and their difficulty with equipment, may encourage them to do their maths in their heads, usually without success. They may also fidget a great deal, because it is a conscious effort to coordinate their body to sit still.

Case study: Andrew Lindsey

Andrew Lindsey is a successful 57-year-old clothes designer and retailer, who lives in Hove, Sussex. He has suffered from acute dyspraxia and moderate dyslexia. One of his pet hates is so-called experts who pontificate on his related disabilities.

'I can assure you, none of these researchers or scientists personally suffers from dyslexia, dyspraxia or ADHD-related symptoms... Only someone like myself, who has suffered relentlessly, can explain the torture that it brought ... all my life.'

He thinks one of the reasons for his problems was that he was born two months prematurely. And despite attending excellent private schools, he was, in his own words, 'physically and mentally very late developing. My parents could never understand why I was so backward and teased so mercilessly at school for being clumsy and thick (that's only the tip of the iceberg)... They just hoped that I was a late developer and I would catch-up, as I got older.

'Sadly, this never happened. To my family's anguish, I left school

---

with no qualifications and continually broke and dropped things. I was a walking liability.'

Andrew came to the conclusion that 'being left handed and very creative, I was destined towards a visual/artistic career choice'.

Despite his artistic flair and some success in the fashion world, he felt that he never fully achieved his true potential because of his disabilities, which also affected his relationship with his girlfriend, Amanda.

Just before ending their relationship, she referred him to a newspaper article about the controversial Dore method. He was immediately hooked and paid to join the programme. This changed his life, Andrew claimed. 'They told me I had a huge brain, which was being totally underutilised.'

After persevering with the method, pioneered by Wynford Dore – which has been criticised by educationalists and scientists – Andrew said he rapidly improved his memory skills. 'And I can now read with confidence without getting tired and losing concentration. I can write fluently, making hardly any spelling mistakes. I now get asked "how do you spell this word and that word?" Would you believe that?

'I can run in the rain without falling over. For the first time in my life I feel safe on my feet.

'My confidence has improved beyond recognition. My ex-girlfriend, Amanda, now tells me she loves me and we are dating again.'

Dyslexia

It has been estimated that there are at least ten times as many academic papers on dyslexia as there are on dyscalculia. Far more research money is spent on the subject and it could be said that dyslexia research is a decade or two ahead of investigations into dyscalculia, especially in the USA. In America, the National Institutes of Health spent $2 million on studying dyscalculia between 2000 and 2011, compared with more than $107 million on dyslexia.

There are many reasons for this. The social taboo about being illiterate is far stronger than being bad at maths. Also, diagnosing dyscalculia appears to be more difficult, although screeners such as Professor Butterworth's are very reliable. Children can be bad at maths for many reasons besides dyscalculia, for example bad teaching. It is so easy to fall behind in maths when central concepts are missed because of poor schooling or absence through ill health. Parents can often compensate by helping with reading skills at home, whereas most find remedial maths far more challenging.

Dyslexia is a neurologically based learning disability that impairs the processing of language. Neuroscientists have shown how reading requires very complex interaction in the brain and dyslexics display different wiring patterns. As with dyscalculia, children with dyslexia usually possess normal intelligence. And, as with maths disabilities, dyslexics tend to fall behind their peers in reading ability. Like dyscalculia it can vary from severe to mild and it does not go away completely as the child matures, but the severity may decline in the case of dyslexia. With dyscalculia, age does not attenuate the condition, though adopting coping mechanisms may ease the problems.

Because of environmental factors, dysfunctional families or poor schools, as well as health problems that lead to stress, dyslexia often goes undiagnosed. The US Department of Health and Human Services, for example, estimated that 15 per cent of students may have dyslexia. But as there is no universal definition or standard diagnostic tool, the statistics may be inaccurate. Although more boys than girls are said to have dyslexia, this may be a gender bias resulting from the probability that boys may act out their learning frustrations in bad behaviour, while many girls may simply retreat into a shell. Although dyslexia and attention-deficit disorders are distinct conditions, around 40 per cent of dyslexics are likely also to be diagnosed with ADD.

Poor reading skills can also relate to hearing and visual problems besides – in layman's terms – the innate wiring problems in the brain related to cognitive skills. Once hearing or sight problems are dealt with (and sometimes they can be quite easily and quickly fixed, with spectacles or hearing aids), the original issue may remain, producing poor spelling skills, as well as reading deficiencies. Dyslexic children have difficulty decoding words, even recognising them as words at all. 'Fixing' dyslexia – with skilled intervention – tends to produce better, faster and longer-lasting results. And further developing coping mechanisms often means that dyslexics are proportionately over-represented in the milieu of visual arts, architecture, music and design. Dyslexics see differently – they often visualise a word or shape from numerous angles – and it can be a major bonus in the creative world, though not very useful for a TV presenter reading from a teleprompt, for example.

Most dyscalculics will come to hate maths, whereas some dyslexics enjoy reading, but simply find it very difficult and time-consuming, compared with the reading speeds of their peers. In both the US and UK, social class is known to intrude on success rates in overcoming the condition. The presence and use of books in most middle- and upper-income families is often one predictor of reading success in later life. And, of course, better-off families may have more time to spend time reading with their children or the ability to pay for extra tuition. Also dyslexia, like dyscalculia, can be self-perpetuating. It is most likely that it is influenced by genetic factors. So mix in inheritance with poor environment and it is probable that dyslexics from disadvantaged backgrounds will need even more help in schools.

Dyslexics may have difficulties with understanding words especially in some ranges of sounds such as *ba, ca* and *ta,* for example. In maths, they may confuse words such as *equals* and *eagles.* Copying the words and repeating them verbally becomes difficult, let alone trying to write them down. As the inner ear acts as a fine tuner for all motor signals (from balance and bodily coordination in general, to rhythm in particular) many dyslexics will have poor balance and spatial coordination, not least because they struggle with complex eye movements needed to coordinate reading. The brain scrambles the signals and the result is that words and lines jump around on the page. The extreme contrast of black text on white background can have the effect of making the letters swirl or blur which is why the British Dyslexia Association advises using dark-coloured text on a light (not white) background to make it more dyslexia friendly.

Many dyslexics find placing sheets of tinted acetate over a page helps to stabilise the text.

Difficulties in hearing the sounds correctly and visually aligning a page obviously make reading far more challenging. A small percentage of dyslexics suffer from light sensitivity. They may dislike fluorescent lighting and often shade with their hands the page they are attempting to decipher. Like dyscalculics, dyslexics may suffer from memory problems, such as trying to recognise visual memory patterns for sound-alike words.

Some key symptoms of dyslexia
♦ Poor sequential memory, not just for words, names days and months, but also for following oral direction
♦ Difficulties with rhyme
♦ Confusion of similar words and also similar numbers such as 6 and 9
♦ Words will be missed, blurred, move around and change size, and so on
♦ So there is a frequent tendency to skip over words or phrases and instead guess
♦ Poor concentration
♦ Writing will be messy, change size and angle and often letters will be in the wrong sequence (just as dyscalculics misplace numbers)
♦ Difficulty learning the alphabet
♦ Slow reading, moving head around and using fingers to track the words
♦ Nuances will usually be missed, written and spoken messages will be taken literally
♦ May not speak well, and may have been slow to be speak as an infant
♦ May be clumsy and poor in working out basic steps such as tying shoe laces
♦ Poor with understanding left and right, before and after and front and back
♦ A small vocabulary is another warning signal, particularly in written work. A dyslexic is inclined to avoid using long words that are difficult to spell.

Responding to dyslexia

Being able to read fluently is such a crucial part of schooling that dyslexics will need the same kind of specialised intervention, patience and one-to-one coaching that dyscalculics require. Some of the environmental factors may be resolved, but the genetic wiring apparatus just will not fit into a conventional school curriculum. If the dyslexia is acute, then a special school for dyslexics may be considered. If it is milder, special educational-needs teachers in a conventional school should work.

♦ Colourful and exciting high-impact books will be needed to arouse interest, perhaps with personal life stories of animals or well-known 'heroes'

♦ Practise at least four to five times a week with sounding words and getting the child to repeat them

♦ Reading practice every day at home and school

♦ In the general mixed-ability classroom a good reader can be paired with the dyslexic. Let the fluent reader go first and let them talk about the story and then let the dyslexic try

♦ Typically, dyslexics will start making wild guesses at words; they may do it less with a paired peer and under reduced pressure

♦ The same principles apply as for teaching dyscalculics – reading should be fun. Use Dr Seuss books with their rhyming stories

♦ For younger children all reading should be done aloud and rhyming songs should be used

♦ Listen to taped books

♦ Multisensory teaching is key for accessing the different learning modalities

♦ For a child with visual difficulties, one traditional solution in a classroom is to place the student in front, near the board/teacher

♦ As ever, allowing more time is crucial.

Dyslexics usually cannot process how words sound – a phonemic deficiency – before the next word comes along. There are a variety of systems, especially in the US, which can help phonological development: Starfall, FastForWord, Lexia, Earobics and Accelerated Reader, for example.

Parents can use these at home and because it is an auditory technique – there are no words on a printed page – it should be much easier to get the children to engage. This helps them to 'sound-spell'. In some cases dyslexics may have a much larger visual, rather than sound, deficit. Coloured paper or transparencies have been known to help, and in some research coloured contact lenses have accelerated learning.

Case study: The 'gift' of dyslexia

The inescapable negatives of dyslexia have been emphasised so far, but the creative possibilities have not. Nearly all the so-called experts on the subject, no matter how well-meaning, can never fully understand how dyslexics see the page, let alone their world. As far as I know, the book you are reading is the first book (apart from memoirs) written by a dyscalculic on the subject of dyscalculia. There is, however, an uplifting book written by a dyslexic on dyslexia: Ronald D Davis's *The Gift of Dyslexia*, with the 's' suitably inverted on the cover (see further reading). Written in comfortably large print and in a direct style, it is a useful self-help book for dyslexics.

Davis establishes the upbeat tone of his book by quoting his experience on a TV show. He had just listed a dozen famous dyslexics.

'Isn't it amazing that all these people could be geniuses in spite of having dyslexia?' said the female interviewer.

Davis says she missed the point. 'Their genius didn't occur in spite of their dyslexia, but because of it.'

Davis argues that:
♦ Dyslexics have vivid imaginations
♦ They can utilise the brain's ability to alter and create perceptions (without drugs)
♦ They are usually highly aware of their environment because

---

they can see it from various angles. (I would add that some may make good combat soldiers because of what the military call 'situational awareness')
♦ They tend to be more curious than their peers
♦ They think mainly in pictures rather than words – hence their success in art and design
♦ They are often highly perceptive and intuitive (despite their often literal interpretation of words and spoken statements)
♦ They tend to think and perceive multi-dimensionally
♦ Davis also claims that dyslexics can experience thought as reality.

Based on his own dyslexic upbringing, he believes that, if the innate positive aspects of the condition are not suppressed by educational or parental authorities, the child will grow up with 'extraordinary' creative abilities. Perhaps in an overstatement, Davis claims that 'the true gift of dyslexia can emerge – the gift of mastery'. He cites Walt Disney as a classic example. The same mental functions that prevent dyslexics from spelling, reading or writing fluently produces a 'knack' for being able to do other, more artistic, things. Dyslexia, it is argued, is the result of a perceptual talent. His book shows how to develop this talent rather that letting it become a liability.

Because of his own dyslexia, Davis is very informative, but he sometimes overstates his case by arguing, for example, that very good teachers rarely have children with ADD in their classes (even though they behave badly in other classes). He also claims, with some verve, that he has helped more than a '1,000 children and adults make their words – and the world – stand still'. Davis deploys a somewhat evangelical American self-help style and he is on firmer ground when he argues that dyslexics – apparently like the ever-quoted Einstein – think in pictures, which, he says, is 400 to 2,000 times faster than verbal thinking. Speed aside, 'picture thinking is more thorough, deeper, and more comprehensive.' As well as more intuitive, he maintains. His main argument is that picture thinking, intuitive and multi-dimensional thought, as well as curiosity, all enhance the dyslexic's mind-power. Davis flies in the face of neuroscience by claiming that dyslexia is a self-created condition, but this claim has to be understood in the context of his conviction that he was at his 'artistic best' when he was at his 'dyslexic worst'. Despite the touches of New Age aspiration, Davis provides a very credible insight into the dyslexic mind. Not least for his description of how dyslexics read:

> ... the mind's eye is mentally circling around the letter as though it were an object in three-dimensional space. It's like a helicopter buzzing around, doing surveillance on a building.

In Britain, Susan Parkinson, an inspiring and eccentric teacher, also argued that dyslexia could be a blessing. She believed she had proved that its three-dimensional vision often generated creative and imaginative prowess, despite the characteristic problems with reading and writing. She helped found the Arts Dyslexia Trust in 1992. She also organised art exhibitions where the work of dyslexic painters, sculptors, designers and architects were displayed.

The dyslexic advantage

Although talk of a 'pathology of superiority' is an exaggeration, more independent research on the advantages of dyslexia has been done.[6] Looking at the negatives while dismissing the advantages has been compared to trying to understand a caterpillar while ignoring that they grow up to be butterflies. That does not mean that dyslexic children should not benefit from school intervention or extra parental support, because the butterfly outcome depends upon not destroying early confidence. Many young dyslexics grow up to reach the top of their fields, from the Beatle John Lennon to the architect Richard Rogers to computer pioneer Bill Hewlett. There are many more examples of less famous super-achievers. Some experts suggest there is a causal link between talent and dyslexia.

Advantages include
♦ three-dimensional spatial reasoning and mechanical ability
♦ the ability to perceive and take advantage of subtle patterns in complex and constantly shifting systems or data sets.

A positive view of dyslexia would change it from a learning disorder to a different learning or processing style. It is a difference not a disability. The dyslexic child's education should recognise this by balancing support for their weaknesses in reading and writing with playing to their strengths in other often more creative endeavours. Encourage and let the child enjoy what he or she can do, as much as focusing on what the child can't do. This will prevent the child's self-esteem being eroded by the school system. One means of securing this is the provision of special classes or even special schools. Without drowning in the academic debate about inclusion versus exclusion – keeping the child with his or her local peers or sending them away to a separate 'special' school – each dyslexic child's needs must be assessed separately. Many enjoy being in a distinct class or school where they don't feel left behind. Other children may have different psychological needs and it may be better to keep them in their local school in standard classes (but with provisions for help out of school hours).

After leaving school, dyslexics should aim for jobs that fit. Generally they should opt for careers that involve problem-solving, trouble-shooting, fixing things or 'telling stories' (public relations, sales, coaching) and most famously entrepreneurship, with Sir Richard Branson and Lord Sugar being prime examples. On the other hand, dyslexics may struggle in more routine jobs that involve adhering to strict principles and rules. Dyslexics are likely to excel in jobs that stress results rather than methods. They will tend to take unconventional approaches. Thus they are more likely to succeed in a smaller firm which permits flexibility rather than a big company with hide-bound regulations. Best of all is to start up a business – that's why so many well-known dyslexics are self-starting business people.

At least working for yourself avoids the problem of disclosure – should dyslexics or dyscalculics admit to their disabilities when applying for jobs? It is hard to answer this because it depends on the job. Personally, I rarely mentioned to any employer that I was useless at maths.[7]

My book is not primarily about dyslexia, but it is important to understand the positive aspects of what some may term 'disabilities'. Frankly, it is harder to make an equally uplifting case for dyscalculia. To take a personal reference: when I was first at the Royal Military Academy (for US readers, the equivalent of West Point) I was always told off for taking shortcuts and being a maverick, if not just a plain nuisance. Perhaps my dyscalculia-influenced behaviour just didn't fit in with strict military parameters. But I continued my involvement with military affairs as a war correspondent and later was requested, on a number of occasions, to return to military life, because my condition had forced me to think outside the box, all the time. Later, I developed useful skills, from operating in Afghanistan with special forces-trained people, to breaking all sorts of rules in working undercover in other war zones.[8] I would argue that, like Davis, a disability, if channelled properly, can lead to a successful, imaginative and often different lifestyle from those not troubled, or blessed if you insist, by neurological disorders.

Readers may ask here how many war reporters, for example, are dyscalculic. The answer is I don't know, and I am pretty sure no one has tried to find out. Another answer may be that because war correspondents often display a wide array of neurologically distinct and eccentric behaviours, they may constitute a complex research zone. I haven't asked my journalism colleagues to try Butterworth's tests, but I have analysed their psychological motivations in my book, *Shooting the Messenger* (see further reading).

So how can positive conclusions such as in Davis's book or my war experiences be applied to our main theme – diagnosing and overcoming dyscalculia?

KEY POINTS
♦ Pure dyscalculia is comparatively rare – it is often accompanied by related conditions such as dyslexia, dyspraxia or attention-deficit disorders
♦ Dyspraxia can be overcome, even sometimes using unconventional therapies
♦ Dyslexia can sometimes be channelled into creative and artistic careers.

4: Testing for dyscalculia

The tests are much the same for adults and children.... except of course the adults have 'times × 2' bigger chairs.

It is important to try to find out whether a person with numerical difficulties has simply a temporary phobia or genuine innate problems. It could be a development issue or the fact that the child has missed a lot of school because of illness. It may be that the child does not like his or her maths teacher, for good or bad reasons.

There are informal and professional means of testing. Obviously if dyscalculia is suspected it is far better to rely on professional and clinical help from special needs teachers and educational psychologists. Once diagnosed, each child will require a detailed bespoke programme, which can last years.

The Dyscalculia Screener
Professor Brian Butterworth, at London's Institute for Cognitive Neuroscience at University College London, has developed a series of basic computer tests that can register an accurate assessment, the

Dyscalculia Screener. What is very useful about these tests is that they depend very little on other skills such as reading or short-term memory or even on a student's previous educational experience. The tests can assess a child's numerical capability independently of their abilities in other areas.

The Dyscalculia Screener uses a standardised software based on three tasks controlled by a computer; it also tests for basic speed in detecting objects appearing on the computer monitor. The three tasks can be divided into two subscales: a 'capacity subscale' and an 'achievement subscale'. The first subscale involves a dot enumeration task and a number comparison task. In these tasks children are asked either to count as fast and as accurately as possible dots appearing on the monitor, or to decide which of two numbers, for example '4' and '7', is the larger. The second subscale involves an addition task where youngsters are simply asked to check if a sum or product is correct, for example are '7 + 9 =16' or '3 x 4 = 9' correct or not.

The software diagnoses developmental dyscalculia on the basis of norms calculated for each year group. Performance in the Screener takes into account both accuracy, namely the number of correct responses, and speed, that is the time taken to produce an answer. These measures are known to be important in the diagnosis of numerical disabilities.

Dyscalculics are also often tested in terms of their general intelligence and other skills. This is to exclude the possibility that problems with numbers may be due to other more general cognitive difficulties. IQ scores are typically calculated by administering standard tests. In addition, reading and attention can also be tested, for instance by using the reading aloud test from the British Ability Scales.

I have mentioned Butterworth's Screener because I am more familiar with his work and because he is a very dedicated dyscalculia researcher. But there are many other ways of measuring. For example, the MALT test – Mathematics Assessment for Learning and Teaching – is used from reception classes to Year 9. The Sandwell method is a standardised test which can be used at the beginning and end of interventions. Numicon not only provides very useful teaching equipment but also a diagnostic assessment tool.

The Jane Emerson handbook

The Dyscalculia Assessment, written by Jane Emerson and Patricia Babtie (with a foreword by Brian Butterworth) is an up-to-date and comprehensive book for teachers concerned with dyscalculia in primary schools.

In the book, Professor Butterworth says that his computer-based screener can distinguish the dyscalculic learners from other lower-attaining learners. 'As with dyslexia,' he says, 'the key to success is making the right assessment and using this to create the right plan for each learner.' The key word is 'each': each child will need individual intervention. One of the major initial points in the Emerson primer is that any assessment works better when there is a significant discrepancy between the child's general intellectual level and their attainment in maths.[9] This can be ascertained by comparing maths ability with other skills such as verbal and literacy abilities.

The Emerson primer advocates an informal friendly assessment, collecting wider information from the teachers, parents and possibly educational psychologists. Then follows a first chat with the child, to find out what they like and dislike about maths. Next step is a structured investigation using the *Dyscalculia Assessment*.[10]

Educational psychologists may also be asked to deploy the Wechsler Intelligence Scale for Children (WISC IV). Besides ascertaining the much-vaunted IQ levels, it can provide information on verbal reasoning, perceptual reasoning, working memory and processing speed.

Parent or teacher – informal tests

No single diagnostic tool is foolproof just as no single intervention strategy is guaranteed. There is no simple 'cure'. If you do not want to use one of Professor Butterworth's tests, for example, you could try more informal assessments. (Specialists do not encourage self-testing, for good professional reasons. Butterworth's tests are not cheap if done as a 'one-off', but informal methods may encourage more use of professional advice.) Teachers and parents can observe their children to see if they have obvious problems with:

♦ Counting
♦ Measuring
♦ Remembering basic maths facts
♦ Understanding shapes
♦ Generally organising information.

Parents and should consider the following list of issues often raised by dyscalculic children themselves:

Listening to the child

❏ I find it hard to copy down on paper a list of numbers when read out by the teacher, or even from the board

❏ I sometimes see a number written down, but when I copy it on paper the numbers may be in the wrong order

❏ I do this with the phone sometimes – even I when I use the number all the time. I have to check and re-check the numbers. I am a bit wary of using a telephone, especially if someone is watching me

❏ I take much longer than other students when I am asked to do basic mental arithmetic

❏ I don't understand what fractions, odd and even numbers, let alone square roots, mean

❏ Even when I sometimes get the hang of these, I have to write them down and carefully work out what an odd or even number is

❏ I can't do my times tables, at least not all of them – I can do the lower ones, sometimes

❏ When I am doing a maths problem in school, even ones I have done before, I forget where I have got to, and have trouble finishing off the problem

❏ When I do manage to complete a problem – and even get it correct! – my working is very untidy and I am still not sure how I worked out the answer

❏ Sometimes when I get it right I can't explain how I did it – the teacher has even accused me of copying someone else's work

❏ Big numbers confuse me

❏ Percentages are very difficult to work out (though many non-dyscalculic adults may also admit to this)

❏ When the teacher says, 'If a man with a tractor can plough two fields in two days, then how long …' I switch off immediately

❏ The 24-hour clock is a mystery to me

❏ I get the symbols confused for division and multiplication sometimes

---

❏ I usually can't remember the terms for the shapes in geometry
❏ I could never work in a shop – I'd spend all day trying to work out the correct change
❏ I have been abroad a few times and can't understand how much foreign currency is in my money
❏ Maths does frighten me
❏ I feel stupid sometimes in the class when everybody else can answer so much more quickly than I can.
❏ I cannot do certain things quickly, like working out the cost of shared snacks in the school canteen, but may be able to if left with sufficient time
❏ My teacher must hate me – she keeps on giving me fractions to do.

If a child ticks at least half these boxes – depending on the child's age of course – it is likely that she or he suffers from dyscalculia.

Stepping in to help should be done as soon as possible, through intervention at key points – in primary school, transition to secondary school and before the exams taken at 16 in the UK.

The child should receive extra help, perhaps private coaching at home or a special teacher in school (see http://news.bbc.co.uk/1/hi/education/8298191.stm for an example of a recent initiative of UK schools to offer tuition to students with low scores in maths). With extra help it may be possible for them to reach an average performance in a year or so, ideally before taking their final GCSE maths exam. Also, they can be issued with a 'statement' (in the UK) that they have a recognised disability and can be allowed extra – crucial – time in exams (although extra time can be allowed even without a statement). The extra time allowance was not designed with dyscalculics in mind and specialists such as Brian Butterworth do not think it helps them very much.

Students can benefit from one-to-one help in class, in small specialised groups also in class, in one-to-one cooperation with a special needs coordinator in school or private coaching at home.

The idea is to use intensive personal coaching so that the child can return in confidence and keep up with their peers in the classroom. Some pupils may need a quick boost of coaching-inspired confidence, while others may need individual assistance throughout their schooling.

Testing adults

As an adult, being certified as a dyscalculic may not tell you anything you didn't know: you are bad at maths. Since the average employer won't know anything about the subject, you are not likely to get any benefits in terms of employment. And since it is not a recognised physical disability (by the Department of Transport) you won't get a handy disabled sticker for your car. But in many cases it can be psychologically satisfying to understand the origins of problems you have suffered from since childhood. It may help some adults to know they are not lazy or stupid – they have very probably a genetic inheritance. In some cases, it may encourage adults to return to some form of part-time education which may address some of their problems, or at the very least to read a book like this perhaps to learn new coping mechanisms.

Diagnosing and remedying adult dyscalculia can be more complicated than with children, though again Professor Butterworth has been working on computer tests. After leaving school we tend to forget our algebra or trigonometry because we rarely use these skills. We do, or should, remember the use of numbers: adding, subtracting etc, and working with decimals and percentages. We need these to cope with money, weights, bank statements, and so on. Adults, especially with maths difficulties, tend to over-rely on calculators. So, without them, when they are tested, the results may not show their true abilities. Refresher courses in adult education can produce surprising results, not least because the motivation is often very high. The brain can be very adaptable (or 'plastic') even in adults. Research has already shown that training programmes can increase function in the brain areas involved in reading; the same is likely to apply to dyscalculia as well.

KEY POINTS

♦ For successful remedial intervention, a professional assessment is vital so a proper programme can be devised for the pupil

♦ Professor Brian Butterworth's screener is a reliable assessment tool

♦ Teachers will find Jane Emerson's primer very useful in the classroom

♦ Initially an informal assessment can be made by listening to comments from pupils with maths problems

♦ Adults may also find an assessment useful.

5: Teaching dyscalculics

Reasoning

'Maths is not about finding a correct answer by following a recipe or about acquiring a page-full of ticks in a school exercise book. It is about making sense of the world, investigating ideas and concepts, making connections, developing and strengthening abstract cognitive skills.'

(Ronit Bird, *Overcoming Difficulties with Number*)

Although I am an experienced teacher, as a chronic dyscalculic I am definitely not an experienced *maths* teacher. So I have consulted extensively with those who have taught maths to dyscalculics. The unique point of this chapter is to approach the subject from the dyscalculic child's perspective. (The child in me still screams when facing maths challenges!) Some of the suggestions will be standard practice for senior teachers and special needs professionals; I do not wish to teach grannies to suck eggs. But please bear with me because there will be some new ideas and angles, I hope, and reminders, not least for less experienced teachers.

Of course a paradox niggles: dyscalculics, like dyslexics, often do not respond well to standard teaching methods – no matter how good – that is why dyscalculics fail in schools. Overcoming the problems often requires high-quality standard teaching PLUS – that is what this book is about.

Helping in this area is not just a case of trying to pass exams. If a child spends his school years feeling stupid, facing failure after failure daily, it would be remarkable if their self-confidence and self-esteem were not damaged, sometimes leading to a range of psychological and employment problems in later life.

For teachers and parents to avoid this, they will need plenty of patience. Dyscalculics' performance can often be very uneven. I said earlier that dyscalculics may perform very well in other subjects and have average or high IQs. In some areas they may be considered suitable for 'Gifted and Talented' sets. Until and unless the condition is diagnosed parents and teachers will often think – 'Is she playing games with me? All her friends think she is smart, etc.'

Even when dyscalculia is diagnosed, uneven performance in maths can be frustrating for all concerned.

The first thing to remember is that dyscalculics cannot be part of a race. Nothing is more discouraging than being beaten to the bell by other kids in class. One way around this is to ask mental arithmetic questions and have the students choose a card with answers from their own desk, and place them face down. They can be checked later.

In other contexts dyscalculics will typically ask classmates to help them, for a while, until the teacher decides the practice is disruptive or their classmates lose patience. When younger, dyscalculics will take instinctive pauses, perhaps by walking around. They need to pace themselves to deal with their specific cognitive issues.

This is common sense – dyscalculics may have problems with time and space. To resolve these problems, they may need extra time and space.[11]

Case study: Providing time and space as well as patience

I consulted an experienced special needs maths teacher, Debbie Segal, who works in Epsom, England. She teaches children aged 7-13, on a one-to-one basis, usually for 30-minute periods, once or twice a week. The students are with her for one or two years.

She defined dyscalculia as 'having no clear understanding of quantity' – a handy definition.

'Most of my kids count with their fingers,' she said.

Don't classmates laugh?

'They do it under the desk.'

She added, 'If they are asked to add 6 and 4, they will start at one on their fingers, count to six, then add 4.'

Her basic technique was straightforward: 'I play a lot of games. They don't think they are coming to me to do maths. It takes away their fear.'

She said she played such games as snakes and ladders. 'That's the most popular. I also prefer to work with money – I have a big bowl of change – they see that as more relevant than using counters, when it comes to other games.

'It's all about inspiring confidence. That's 100 per cent of what I do. That's why I play games. It takes half a term, or even a term, to get their confidence back, at least within my lessons.

'I give them confidence with the maths they are comfortable with, the maths they know.'

They approach maths in a different way. 'It's all about small steps so they never lose confidence. I can tell in their body language when they start to get uncomfortable and they are losing confidence.'

Debbie Segal said that her main aims were:
♦ Confidence building
♦ To do maths at their own ability
♦ To make maths 'concrete'.

'Children with dyscalculia will always need concrete examples. Many still use fingers or a calculator, because *they find it difficult to visualise quantity.*[12]

'You have to give them a good strategy – they need to find it out themselves – you can do that if you understand how their minds work.' She added, 'After 13 quite a few go back to the mainstream for maths.'

She did caution, however, that 'Dyscalculics will always need support.'

The essence of her approach is: you have to make it fun, so that it takes away the importance of speed in calculating. 'It doesn't matter how long it takes to get to the answer – it's not a race.'

Her one-to-one teaching obviously involves individual differentiated teaching, which is not always possible in a large classroom scenario.

She defined her method as 'multisensory'.

So what does that mean?

Debbie Segal listed her reply:
♦ Verbally
♦ Visually
♦ Kinaesthetically.

'That is deploying the child's senses – especially touching things, for example, by using a recipe to make cakes. They need to, and will, understand quantities, not least because they want it to taste good when they eat it.

'This is far less competitive, pressure-free,' she argued.

She also maintained that dyscalculics are on average two to three years behind the mainstream – in maths. 'But they can catch up *one year per year* with good coaching,' she said firmly.

Debbie Segal had her students for at least a year – now they all *like* maths, she said.

'That's what makes my job worthwhile. To be a phobia-buster – that's my aim. My games show that maths can be fun. It doesn't have to be boring and terrible. I try to show that it's relevant.'

Helping to teach dyscalculics

Individualised tutoring is a key factor. Dyscalculics will normally need regular one-to-one tuition, in school or at home. Specialist dyscalculia support is sporadically available from local educational authorities in the UK, however.

General principles in this book will be applied by many experienced teachers; a few of them may need highlighting:

♦ Use concrete materials, from Numicon to Cuisenaire rods, and get the children to engage with them. Concrete always trumps abstract, especially for younger children

♦ So use as many visual aids as possible

♦ Depending on the age, play lots of card and board games

♦ Talk about the methods you are using

♦ Think about short, snappy and efficient methods – the fewer steps the better

♦ Concentrate on the journey to the solution rather than getting the right answer first time; the pupils can learn from mistakes

♦ Encourage the child to develop his or her own strategies

♦ Develop key core techniques – strategies that can be applied to a range of problems

♦ Build on what is already known: use familiar methods
♦ Consolidate through regular repetition. 'Overlearning' aids long-term memory.

Specific comments for teachers of dyscalculics
♦ Be sensitive to apparent disruptive behaviour if they ask classmates for help. You need to take control of that, especially in a full classroom. Sit the pupil nearer to the teacher so that reminders can be given more easily, and more discreetly, without highlighting the dyscalculic's slow pace. Obviously a support teacher will make it easier in a large class
♦ Tasks will be set in simple steps, and accept that their completion will take much longer than average
♦ Check written work frequently so that errors can be explained while the pupil remembers what the original task was. The student may well appreciate instant answers and the chance to do it again once their mistakes are explained. They will often misperceive what the problem is
♦ Homework tasks must be set out much more clearly than normal
♦ Use calculators to help explain the decimal point, and compare with money values, which always motivates students. Relating to real-life situations is important. Calculators can give confidence[13]
♦ Compare and contrast written work with any oral summary they might give – they might get the task right, but may misread aloud what they have written, because of confusion with interpreting symbols orally. This may be a language issue, especially if the student is also dyslexic. Many students find it hard to extract the maths from word problems. When mixing the two subjects of English and numbers, the different disciplines get transformed into an unintelligible language called 'Numglish'
♦ On the other hand, if the student has strong auditory skills, this strength may give them extra confidence, get them to read out loud the tasks, and listen carefully.

Depending on whether primary or secondary pupils, lots of visual aids should be deployed. The key issue is to get the students to 'visualise' maths.

♦ Allow the students to count with their fingers

♦ Use mnemonic devices to learn steps of a maths concept. One eight-year-old pupil said she used 'riddles' to help with her times tables: 'One of my favourites is "I ate and ate until I was sick on the floor. 8 x 8 is 64."'

♦ Use rhythm and music to teach maths facts and to set steps to a beat. They will need extra time to memorise maths facts. Repetition, and rhythm, may be very helpful

♦ Many children in the US especially have benefited from 'mathematics and movement' programmes, where they hop, skip, dance and count at the same time[14]

♦ Use standard squared paper and encourage them to keep the numbers in line

♦ Different background paper may help – think of using yellow-coloured paper when photocopying

♦ Perhaps use triangular or easy-grip pens and easy-to-hold rulers

♦ Avoid clutter, especially on tests – allow scrap paper with lines and ample room for uncluttered computation; though the final work should be written in the school book as part of the solution. (Also, scrap paper is not allowed in the UK GCSE exams; if a student has used scrap it has to be handed in. Even if the final answer is incorrect, marks will be given for sound working methods.)

♦ Use charts where possible

♦ Suggest use of coloured pencils or highlighters to differentiate problems (perhaps red for addition and blue for subtraction)

♦ Use diagrams and draw maths concepts

♦ Make sure that the student has sufficient computer time scheduled for the exercises

♦ If possible, calculators or computers with speech-synthesis functions can be used. They can enable the user to select options to speak and at the same time display numbers, functions, equations and results. At the every least, deploy a large calculator with big number buttons of different colours.

Patience again is the ultimate virtue. A solid grasp of the basics is essential to enable a student to understand concepts at a higher level. The slightest break in logic can overwhelm the student and cause emotional upsets. Students may appear to fully understand the tasks at hand, and then get every problem incorrect in a test, especially in a full class. One to one, five minutes later, according to interviews with some teachers, they can get them nearly all correct by working with the teacher on a whiteboard or mini-whiteboard.

Teaching dyscalculics at primary level

General comments

Dyscalculia is usually very identifiable at primary level. As I mentioned earlier, without corrective coaching, pupils may reach their peak at around eleven years of age, and then perhaps progress just *one year* in secondary school. At primary level, the average nine-year-old dyscalculic may have the ability of a six-year-old. This gap will widen at secondary level, without proper remedial coaching and regular intervention. Otherwise, at 16, a dyscalculic sitting for his GCSEs may not have advanced much further than primary-level standard.

Some may be able to count to high numbers, but will have no concept of their relative position on a number line. They have no useful starting point to begin calculating, except to start counting from one, usually with their fingers. They usually don't understand that sequential whole numbers have the same intervals between them, so that 8-10 may seem a bigger gap than, say, 40-42. (A sound background in the use of, say, Cuisenaire rods may help to alleviate this.) Using a book may help. If the pupil is not dyslexic and has a favourite book, perhaps the proverbial Harry Potter, ask them to bring it in to school. Then ask them to find page 230, for example, and then ask them go back and 'find the bit about, say, the dragon, on page 22, I think'. This is a concrete method of getting the child to go up and down the number sequence, with a useful purpose, for them.

With younger pupils, the aim is to enthuse them with the idea that maths can be fun. And so is 'doable'. Psychologically, the emphasis on maths skills for academic progress and job opportunities should wait until later in secondary school. As with the Jesuit boast – 'Give me a child until he is seven and I will give you the man' – so too in teaching dyscalculics to cope. Diagnose early, intervene effectively and avoid the elaborate construction of phobias that can swamp teenage development. The longer the sufferer fails to 'manage' his or her own dyscalculia, the more difficult it will be to reach an 'average' maths standard.

Be Careful with language
Children will get confused. If you use 'take away' for 'subtract', the child may think of pizzas!

Children can be quite literal, especially if they are dyscalculic. Little Johnny was proud that he could count from one to ten. When he was asked to do it backwards, he got off his chair and walked backwards while counting forward to ten.

Being literal, and logical!
One exam question asked: A hat costs £5 in the UK. The same hat in Germany costs 8 euros. The exchange rate is £1 = 1.40 euros. Where would it be cheaper to buy the hat?

Answer: 'Taking the plane into account, the UK.'

If a child has hearing issues or second-language 'interference' (as it used to be called), then things can sound differently. 'Equals' can be interpreted as 'eagles'. Perhaps don't use 'equals' and say 'fair share' – help the dyscalculics to understand that halves have to be equal in number – make an obvious comparison with football fields being two same-sized halves and what half time means. You could relate to the idea of 'fairness' in games.

So be slow when using synonyms – 'add' and 'plus', or 'subtract' and 'take away'. Keep vocabulary single, simple and consistent at first. (Consistency may of course be a problem if the pupil is subjected to changes in teaching staff.) It took me years, for example, to understand they meant the same thing. So simple consistency at first can be a 'plus' factor. Information cards with symbols for the maths operations with the common terms for each are helpful.

Also with multiplication: when I heard the term 'times', I thought intuitively of minutes and hours. I didn't have visual-spatial problems so I didn't confuse + and x, but some dyscalculics will.

Most dyscalculics will have difficulty talking about maths, using the vocabulary, because they – we – can't generalise or see the patterns, let alone describe them. Learning the basic facts is like learning to spell, but I didn't always see the logic. I thought fractions were counter-intuitive, no matter how they were explained. One-hundredth seemed larger than one-tenth and likewise .007 seemed larger than 7. But when decimal money (not then used in the UK) was introduced it made more sense.

Memory issues

Dyscalculics often have short-term working-memory problems as well as long-term memory deficits. (I am referring to memory of numbers deficit; their memory for other materials is rarely tested.) Pupils will often forget the question, or get lost in mid-calculation. They may forget recently learned facts very quickly. They may resort to deploying the whole times tables – if they have mastered that. Longer-term memory problems can relate to failure to remember basic maths facts from times tables to the meaning of symbols for multiplication and division. We return again to best practice: rote-learning and regular testing.[15]

Slowness

Dyscalculics may well be impulsive, inattentive and sometimes disruptive learners. They often fail to learn by trial and error – they will sometimes make the same mistake over and over again.

Generally, don't move on until the child has mastered the project of the day or week, or month – or maybe in some cases – a year. Build confidence. For example, if the child is weak in mental arithmetic, let him use a calculator. Or a computer. Deploy the copy-and-paste function, for example with patterns of multiple of tens – pictures of ten small cars or other attractive shapes such as smiley faces. (Experienced teachers will be wary sometimes of using electronic calculation. It depends on the aim: numeric competence or understanding of mathematical concepts.)

Reliance on 'physical' counting

Let him or her use fingers, counters or Numicon, if they continue to struggle with mental arithmetic and can't estimate theoretically. Children need to interact with the multilink cubes and the number lines they may be using.

Spatial problems

Their layout will usually be untidy, and they often cannot copy correctly. Many will confuse, say, 12 and 21, or the position of the decimal. Space out the worksheets – use two or three pages, if you normally use one. The 'compression' of so many figures is usually daunting for dyscalculic children. Encourage them to use one sheet of paper for each question if necessary; though shorter – sweeter – questions might be the optimal solution. (Teachers might also make a practical objection about the waste of paper. Extra space might also encourage untidy working: pupils may increase the size of their writing to fill the space.)

Telling time on analogue clocks is often one manifestation of both maths and spatial problems for dyscalculics. Parents and teachers should encourage the children by regularly associating keenly anticipated events – such as favourite play activities, breaks in school, visits to friends or favourite TV programmes at home – with the use and understanding of the clock. It helps at home if the clocks are large.

As one senior maths teacher told me: 'You'd be amazed how many teenagers cannot tell the time. They just use mobiles.'

Although eventually I learned to tell the time, I have spent my whole life underestimating how long tasks take, whether travelling a relatively short distance to a dinner party or the time it takes to write this book, for example. I am always over-optimistic and usually get it wrong, except thankfully for timing on live broadcasts when adrenaline seems to act as an extra discipline. Most of my friends tell me to arrive 15-30 minutes earlier than the real time. Parents often do the same or even alter the clock by 15 minutes.

Different brains
Patience from the teacher and practice for the child have to be combined. Although some scientists suggest that dyscalculic children are genetically wired differently there is no point in being fatalistic. The brain is surprisingly plastic. It is not a muscle and the gym parallel is overdone, but constant work-outs with maths can aid mental fitness and build up skills. Neuro-imaging research suggests that cognitive activity can lead to changes in brain structure and function.

Some children are described as possessing the 'inchworm' and others the 'grasshopper' type of logic. One plods mechanically and uses many workings, while the other can often intuitively see the wider picture and estimate correctly without detailed notes. The inchworm will seek set formulae, will not estimate and rarely verify his conclusions. The grasshopper will explore, can work back and adjust, and will verify his or her results. A few dyscalculics, and many dyslexics, can display highly imaginative and complex leaps in logic. Usually, however, dyscalculics tend to be slow-moving inchworms.

Each child will have different and often distinct variations of dyscalculia – there are as many as 50 discrete symptoms, let alone individual speeds. (See appendices for complete list of symptoms.)

Making connections

Teaching maths is largely about showing patterns, but it's also about connections, both to other numbers and to real life. The ideal – even for dyscalculics – is to make maths fun. Maths is a study of pretty patterns and, in theory, we should all love patterns.

Maths, of course, must never be just about tests on Wednesday mornings. Questions should be asked regularly. How many of you are taking part in the play? How many people are going to your birthday party this weekend?

Allowing for the weather, extend the classroom into the playground, or the sports fields (and home). For example, get a child to choose a team in five-a-side football. In the gym, ask them how many benches there are. How many can be bought if each costs x?

In the class get them to discuss a visit to see their grandparents:
♦ How long were you there for?
♦ How long did the journey take?
♦ How many miles was it?
♦ How old are your grandparents?
♦ What years were they born in?
♦ How many grandchildren do they have, including you?

This is an example of how to also involve parents in exercises.

Use a calendar to discuss:
♦ Birthdays
♦ Parents' birthdays
♦ Birthdays of siblings
♦ The dates of Christmas, Easter, Ramadan etc.
♦ Special events – school carnival. When is the play or sports day?

So they can sense where they are in the cycle of the year.

Good maths students will see the abstract patterns, but the dyscalculics will not – hence the need for the 'human interest factor' – visiting the zoo or seeing granny. If you are not dyscalculic it is hard to understand that others cannot 'read' the 12 hours on an analogue clock, so imagining the shape and pattern of a whole year can be very daunting. (Actually, the analogue clock is very complicated. For example, 8 can mean 8 o'clock, 20 o'clock, 40 past and 20 to.)

Verbalising
Verbal and audio talents may be much better than children's visual-spatial skills. The teacher should carefully speak aloud the questions, as well as displaying the written form. Also get the students to calculate out loud so the teacher can track their thinking. Obviously, verbal praise is important as well. It may encourage the child to think strategically and not just number

crunch. Talking to children is good when a student learns better by listening; and talking aloud may work well for pupils, too.

You can get pupils to write on the board and talk through their working out. Most like working on the board. Another successful strategy (loved by Ofsted inspectors *et al*) is the use of probing questions. Get the (correct) answer to a verbal question. After a suitable 'well done', then ask, 'Can you tell me a little bit more about how you did that?'

Reducing stress
For the dyscalculic, maths – and the fear of failure, 'learned helplessness' – can cause anxiety, stress and even panic attacks.

Again understanding, patience and one-to-one coaching can prevent:
♦ Fear of asking for help, and being told to listen better or 'work it out yourself', so from the start the student feels defeated
♦ Not even knowing how to start
♦ Concern about bothering the friend in the next seat – or provoking the other kids into calling him/her 'thick'. One strategy is to arrange a 'working buddy'.

Otherwise this leads to:
♦ Fear of failure – again
♦ Fear of parents' anger – again ('Tiger mothers' are not helpful in this context.)

Again the patient initial application of standard teaching methods is also required:

Teacher guides the individual child – they practise and at first the prompts are very direct – but they fade away as the child gains confidence and skills. They are not left on their own to fail – to themselves or their peers (though fellow pupils can often be more patient than teachers and they use relevant 'kiddy-speak').

Setting work
For class or homework:

Space out the work – much bigger spaces for dyscalculics – without necessarily emphasising this fact to their peers. (Say: 'Anyone who needs more space for workings is free to take as much space as they require…')

Too many questions packed together can be psychologically disturbing for the average dyscalculic, though it is fair to say that all pupils dislike cramped and busy worksheets.

Use extra pages and extra space.

Laying down the law
Yes, maths is about learning some facts and remembering them; for example, working from left to right, except when you go from right to left when working on calculations written vertically.

Once moving to vertical calculations, they have to move right to left to deal with the units before the tens.

$$\begin{array}{r} 150 \\ -27 \\ \hline 123 \end{array}$$

Be phonetic: generally you say the numbers in the order they are written BUT in the case of 13-19 the last key number word is spoken first. This may confuse dyscalculics – when you say fifteen and write 15. The 5 comes before the 1 denoting 10. The wording of eleven to nineteen is actually not logical in English, compared with most other languages. (Chinese, for example, and languages that borrow their number word system, such as Japanese. But German is as bad as English: *elf, zwolf, dreizehn, vierzehn* – 11, 12, 13, 14 and gets worse for the next decade: *ein und zwanzig* – 21.)

Wall charts – often used at primary level – are useful. The whole 1-12 chart (or 1-10) giving the composition and decomposition of numbers is very handy for the dyscalculics to look at when they are doing their workings. A quantity chart is also useful, demonstrating a comparison of a teaspoon up to a quantity to fill a petrol can.

Mental arithmetic

Most dyscalulics will feel overwhelmed by the very thought of mental computation. Let them use calculators, or workings on paper with the concepts they have mastered. Giving confidence is the key. Quick-fire mental arithmetic problems in a class are the dyscalculic equivalent of being asked to stand in front of a firing squad.

Holding them back

A balance needs to be struck with mixed-ability classes. A slow pace with dyscalculics gives them firm foundations. Paradoxically, in some cases, narrowing and slowing the curriculum to their areas of weakness – in the hope that they will master more number facts – may sometimes be counterproductive. Sometimes by broadening their curriculum they may find areas they can do and so enjoy them. In my own case with geometry: I didn't actually enjoy it, but I could usually understand it. Keeping students strictly bound to narrow areas can be frustrating, and as bad as rushing on to brand new concepts.

They could, on a computer or in a booklet, create their own 'Useful Facts' book to act as an aide memoire – they can record information they have trouble remembering (number facts, examples of workings, or rhymes, rules of games etc).

Rote learning?
Learning by heart can be very useful for mental arithmetic, reciting rhythmically or singing in class.

Sometimes it can work, especially with times tables. But, as we all know, mechanical recitation does not necessarily equate to understanding. One way of checking this is to give an answer and then try to get the student to work out the question – a sort of detective game (though kids are rarely misled by such ruses after a year or two at primary school). Of course, experienced teachers will often invert questions to test understanding: 'The area of a square is 64m^2. How long is each side?'

Guessing
Estimating is an important exercise for developing understanding of amounts. To get away from mechanical rote learning get them to do estimating games with everyday objects – how long is this scarf? – how heavy is this pile of books? Each child guesses and then praise the one who is closest after a demonstration of the precise measuring. Then continue. This belt is so long, how long is this piece of string? Pupils can play with coloured water – guessing volumes – so long as you don't mind the mess.

'Tell me and I'll forget.
Show me and I may remember.
Involve me and I'll understand.'
Chinese proverb.

Teaching older dyscalculics

The same principles must be applied – don't rush. Be patient.

Memory issues
Dyscalculics will have problems with short-term and often long-term memory, as well as active working memory. Don't overload the student – and you may have to break all the rules at least in education in England. Be bold: ignore the syllabus, if it demands tight schedules, and disrupts the slower learners (assuming extra tuition or private coaching are not possible). You may need to dedicate a teaching assistant to this role.

Use small chunks of maths information – as often as possible – to get into the short-term memory which may eventually be filed in the long-term memory. The size of useful – working – memory can be developed and enlarged as long as the student feels confident and is not rushed.[16]

Apply the following common-sense rules:
♦ Ask whether the dyscalculic understands the maths terms used
♦ Use bite-sized information, one bite at a time
♦ Use rhymes, jingles and music
♦ Deploy lots of coloured charts
♦ Use big post-it notes (and even a large, lottery-winner-size cheque on occasion)
♦ Give an individual check list for the session
♦ Always allow extra time.

The more we learn, the more we remember. And the more we remember, the more we learn.

If you are teaching in a busy classroom, and not one-to-one, and without a teaching assistant, all the above need a great deal of teacher planning. Just as each teacher will have his or her own teaching style, so each student, especially a dyscalculic, will have his or her own learning style.

Learning styles

I touched earlier on the differences between 'inchworm' versus 'grasshopper' learning styles. The inchworm will be formulaic, will work down a column and will rarely estimate. The grasshopper is much more intuitive, will work inside his or her head, less on paper, and will often estimate accurately and quickly. To repeat: the dyscalculic will nearly always stand in the inchworm ranks.

Students will learn according to their instinctive styles.

Auditory preferences. Some will prefer verbal instructions (often as reinforcement for written exercises). They will be more likely to be talkative and often mutter to themselves while working. They are more likely to remember initially what they hear. Such students may work better, in their own slow time, with questions on audio players/dictaphones or audio recorders in smart phones (if they are allowed in the classroom). They can play back the questions as often as they want at home.

Or a teacher can record short tests with the student and talk through their approach and then play it back. If the child gets the strategy right, he will probably remember it for a long time. The same when the teacher explains what went wrong – and then allow the student to get it right a second or third time. This will play to his auditory strength and reinforce his longer-term memory. The recordings could also be kept as part of a record of the student's progress. Students with auditory strengths are often disinclined to take notes; particularly at secondary and tertiary levels, they should be allowed/encouraged to record lessons.

The teacher can also get the pupils to talk to each other to discuss methods and probable answers. This will need to be structured to avoid a babble of anarchy.

Visual preferences. They remember what they see. They may well be doodlers and learn best by pictures. One senior teacher reported excellent results with 14-15-year-old dyscalculics who each had access to a computer and were given the chance to play darts games on it. The coloured dart-board seemed to make a lot of difference.

Visual learners obviously prefer to see things – colourful displays written on a board or video clips. They could use coloured pens and pencils for making notes. They would respond positively to using mind-mapping skills for learning or revising maths skills.[17]

Kinaesthetic leaners: the touchers/fidgeters. Often the more hyperactive student will learn best through three-dimensional equipment (though perhaps playing actual darts in the classroom would offend the health and safety brigades). This is perhaps best taught with apparatus such as Cuisenaire rods, counters, dice, etc to stimulate learning of maths in a practical way. Setting up a pretend shop is a fun way of teaching a variety of concepts: the four rules of number (add, subtract, multiply, divide) with plastic money; percentages to calculate discounts; weights and measures using scales; and fractions and division by dividing a cake into halves, quarters, eighths, etc. Money also simplifies the teaching of decimals.

All good teaching relies on all the main senses. The secret of good dyscalculic intervention is to focus on the student's instinctive preference. In what way do they first solve problems – listening, talking, drawing, writing, visualising or moving things around?

Teaching short cuts

Sometimes dyscalculics will find their own methods of getting the right answers – and they may be long-winded. Teachers should let the child continue so as not to undermine their confidence – even if the rest of the class has absorbed the shorter, faster method. Ideally, come back to the shorter method individually with the dyscalculic and, often, the child will adopt the quicker method, explaining that he has decided himself to use the shorter route. Thus the dyscalculic gets to the right place under his or her own steam and in their own time.

Getting lost

I have noted earlier that dyscalculics – especially when they move to secondary school – may tend to get lost. Spatial disorientation is a classic symptom of dyscalculia. Adult dyscalculics may spend their whole lives not understanding the difference between left and right, let alone north and south, even when they become otherwise proficient drivers. Social media sites for dyscalculics often mention the extra anxiety that dyscalculic learner drivers face. But being spatially disoriented does not mean that the person is dyscalculic.

...being spacially disorientated does not mean that the person is dyscalculic.

Older pupils will typically
♦ Confuse west/left and right/east
♦ Confuse clockwise and anti-clockwise.

This will lead to all sorts of problems with measuring shape and space. Students also have to learn a complex vocabulary – horizontal, vertical, isosceles and so on. They may have trouble reading the terms let alone understanding them and then using the them in exercises. One way of dealing with this is to take the pupils outdoors.

Case study: Working outside

Jenny Muddiman, an experienced senior maths teacher in Surrey, England, had been a PE teacher before she also took a maths degree, and would often use active outside methods, whether with Duke of Edinburgh awards on weekends, or helping with maths problems in school.

'Loci. It's a hard topic for pupils – it's quite difficult,' she told me.

I thought: it's a hard topic for most adults as well.

'You can be practical if you go outside. To demonstrate it's not just theoretical. It will also appeal to kinaesthetic learners – people who like to learn by doing. Also it'll appeal to visual learners. And kids who like to be active.'

How do you define 'loci'?

'Loci is a set of points which obey a rule. For example, an area where a mobile phone signal can be picked up. Or where the beams of a lighthouse can be seen. Or not planting grass seeds within two metres of a tree.'

'One child is the tree. The other pupils have to stand around him/her and obey the rules of not being less than two metres.'

Jenny Muddiman added: 'You could say one child has the dreaded lurgy: you need to be more than a metre but close enough to hear them.

'For trigonometry, I use a clinometer. The kids love it because it looks like a gun. It is for measuring an angle of elevation. I usually give out one clinometer between two students. You also need a surveyor's tape measure (up to 30 metres).

'Choose a tree – or anything high. They measure ten metres from the tree trunk. Then they use the clinometer to measure the angle of elevation. Then you work out the height of the tree using a trig ratio. They can do buildings as well. You can count the layers of bricks and then work out the height of the building or wall by calculation once back in the classroom.

'They love going out and measuring; they hate going back in. This helps to show them that maths is fun. Of course you need a nice day, and trees that are away from windows. Otherwise all the other students will say to teachers indoors. "When are we going to do that?"'

Jenny Muddiman also explained how she uses compasses outdoors.

'To teach them bearings, you can create a treasure hunt. Give them written clues where they have to take bearings from specific points to a feature. For example, "You walk on a bearing of 075 (degrees) for 20 paces and where are you?" You need to be in a place where are lots of features, not an open field.

'You give them bearings and distances so that they can complete a course. You could put letters on targets – which they can rearrange as a clue. They need to start in different places – so clues should vary.

'The essential skill learned is to understand a bearing. Another practical skills exercise, especially for younger kids, is to use trundle wheels. You can measure netball courts, for example, or the length of a path. "How long is the path from Mrs Smith's classroom to the main entrance?"'

'They love anything where they are collecting their own data. For the dyscalculic kids what you are doing is adding creativity and excitement, rather than routine written exercises. They have used "multiple intelligence" – verbal, written and listening skills. I consolidate the external exercises when we get back to the classroom.'

Jenny Muddiman added: 'There is a huge value in their describing their methods to other pupils – it gives them confidence, especially if the dyscalculic child has good verbal skills. "You solve the problem and then tell your mates – and see which method is best." Get your pupils to talk about what they are doing; this is very valuable. In a classroom situation get the students to chat in pairs. To avoid idle chatter, give them a short amount of time and the expectation they will feed back to the rest of the group.

'For example, I give my class a task to create a poster [A4 or the bigger A3] describing their method for another group of pupils to follow. They will use child speak and then swap over and get the pupils to evaluate how effective their methods poster is.'

An important component of teaching is cooperation with parents. Teachers could compile a short questionnaire, especially if the child is suspected of having dyscalculia. It can provide useful information as well as generating a series of joint strategies (discussed in the next chapter).

KEY POINTS
♦ Teaching dyscalculics is never part of a race. They need extra time and mental space
♦ The importance of multi-sensory teaching
♦ Individual programmes are needed, ideally one to one
♦ For primary-age children, be careful with the use of language and avoid synonyms
♦ Mental arithmetic tests are like standing in front of a firing squad for dyscalculics
♦ Just as each teacher has his or her own teaching style, then each dyscalculic child will have his or her own learning style
♦ Outdoor teaching can put creative fun into teaching methods.

Mental arithmetic tests
are like standing in front
of a firing squad
for dyscalculics.

6: For parents of dyscalculics

As part of the partnership between the school and parents, a teacher who suspects that one of their pupils might be dyscalculic could send a questionnaire to the parents.

Questionnaires for parents of children with possible maths difficulties

A. Pre-school/nursery
  1 Did your child apply counting skills, say to groups of toys?
  2 Did they touch and count items one by one?
  3 Can they remember their numbers from one to ten?
  4 How good is their memory, for example learning nursery rhymes?
  5 Have they learned the days of the week and months of the year?[18]

---

6 Do they know what today's date is? (Although many 'normal' adults will fail this test!)

7 Do they take a long time to work out which is the bigger or smaller of two objects as well as numbers?

B. Initial primary

1 Did they use their fingers to count longer than their peer group?

2 Have they shown difficulty with writing numbers, for example reversing digits or confusing numbers such as 12 and 21?[19]

3 Has their numeracy generally developed more slowly than their fellow pupils?

4 What is their general attitude to maths? Do they try to avoid it or show signs of anxiety?

5 Does your child forget what type of sum he or she is doing while in the middle of doing it (for example, switching from subtraction to addition)?

C. Older children

1 Is your child generally anxious about maths?

2 Does maths homework cause him or her stress, as well as cause stress in the family?

3 Does your child sometimes/often just give up on maths homework?

4 How do you persuade him or her to continue with their maths?

5 Or does your child often ask for help from you or a sibling?

6 Are you sometimes unsure how to help?

7 If you spend time helping your child – how much time per week?

8 Does a parent have a special interest or competence in maths?

9 Have you considered extra coaching, beyond what the school is able to provide?

10 Would you be interested in co-operating more with the school on developing joint exercises for your child?

Many of the above issues may present themselves in children who are not dyscalculic, but this is a general pointer to the problem especially if an aggregate of maths-related anxieties persist.

Quick check for parents
To test your child's counting skills ask him or her to count out loud forwards from 1 to 30. Next ask your child to count backwards from 30 to 1.

7-11 years old: Ask them to write down the numbers 1 to 50.

12-14 years old: Write down the numbers 1 to 100.

Slowness or errors with the above is a typical sign of a maths difficulty.

Education is impoverished if the child thinks mathematically – usually reluctantly and unproductively – only in the classroom. They need to comprehend – and begin to feel at home in the world of maths – by using the language and logic of numbers instinctively at home as well.

Pre-school
Educational psychologists always tell us that the first five years are crucial in the intellectual development of a child. You may not want to ban TV or computer games, but you can spend time developing a number sense in the child. Count the houses or lamp-posts as you walk past. Talk about numbers inscribed on houses and gates. Ask your children to work out how many apples you have bought in the shop or how many toys his or her brother received for his birthday. When you use picture books ask questions about, for example, the animals: which is the biggest tiger? Make or buy coloured cards and bricks with small numbers on and lay them out in correct order. Teachers use the phrase 'guided practice': you guide and the children practise. Introduce them to games such as snakes and ladders as soon as you can. The old-fashioned simple games are often the most effective.

By the time they get to school they should have some ability with size and numbers, even if it is only up to five … if the child is dyscalculic. They need to feel some familiarity with numbers before they go to school.

I was lucky: I could read before I went to school and that gave me a headstart and the confidence to overcome, or sidestep, my maths deficiencies.

School

Teachers and parents should cooperate closely in dealing with maths work. Parents should try to ensure that homework, and any parental help, is done in the early evening, otherwise pupils may indulge in evasion tactics. In developing a good homework habit, try to evolve the notion of a contract. Homework, it should be explained, is not something that school imposes just for the sake of it. It is a means for the school, parents and child to learn in general and overcome the dyscalculia in particular. In the contractual partnership the child organises and does the homework, while the parent supports and praises (and provides occasional bribes/rewards).

Agree on a suitable regular comfortable place to do the homework, not on the floor or in bed. A kitchen table may work, despite distractions, because it allows the parent to keep an eye on things. (Working in a bedroom may degenerate into endless distractions on a mobile phone or other electronic gadgets.) Start homework within 30 minutes of returning home from school. A fixed time won't work usually because of traffic delays or after-school clubs etc. No television, computer games or other relaxations should be allowed until the homework is complete. Maybe get them to stay in school clothes, to burnish the concept that homework is part of the organised learning process.

Ask how long the homework project will take and allow reasonable mini-breaks if the study requires a lengthy session. Praise them for completing the homework, on time; and keeping their homework diary up to date. Alternatively, put up a weekly homework organisation chart on the kitchen notice-board. It would have subject, date given, work due in, work done early, mark/comment etc. A thorough parent-child partnership could also create revision charts for exams. If all else fails, get them tied into a homework club at school (if there is one). This may reduce stress at home.

Prise them out of the misconception that maths is just a boring or stressful element of schoollife. At home and when out and about show them that maths is a vital and fun part of everyday life.

Maths, you can explain, keeps everything working, not just the plethora of gadgets attuned to the internet. Maths keeps aircraft in the sky. It allows Formula One drivers to go extra fast and it stops gigantic skyscrapers from falling over. Maths drives opinion polls. It controls traffic lights, helps to (or should) get crowds safely in and out of massive sports stadiums and helps to design the glasses the children may be wearing. Maths may lurk behind the scenes, but it is everywhere.

> ### 'Numbers are the handwriting of God'
> – *scientist in the film Pacific Rim.*

Some parents may lack basic educational skills themselves, especially in maths. One recourse might be for one or both parents to take up adult education themselves, separately, or contrive to 'join in' with any coaching they are paying for at home. They could also share in some of the useful maths games on the internet such as The Number Race, designed for primary children, or the Kumon programmes. Try also the games on http://low-numeracy.ning.com

Some schools offer 'maths for parents' sessions. These will typically target the number work and methods being used in the respective schools. If there is a series of sessions, parents should be encouraged to suggest what topics they want to do in the following week.

Parents should have a clear idea of what homework is being set and when, perhaps keeping their own diary. This will help to make sure correct books are carried back and forth to school, and what needs to be kept at home. The pupil could be urged to make a home copy of his or her special maths facts book, which he or she may be using at school. If the parents are mathematically adept – a big if these days – then parents can reinforce the lessons taught in school, using a digital recorder perhaps if the child prefers a more auditory approach.

The school may also have arranged access to websites such as 'Mymaths', a subscription-only site, or BBC Bitesize.

Parents will know not only the email of the pupil's class teacher/tutor, but also the school website page plus the passwords required to access information such as current homework. And parents involved in the traditional school run will have the phone number of a parent or classmate who will know what homework Ali or Susie has to do tonight. Dyscalculics will often be masters at evasion so don't let them get away with it.

Even if your dyscalculic child is not a manipulative maverick, they will tend to be disorganised and forgetful. They will have an innate tendency to misplace books, be late for lessons or even forget – genuinely – extra tuition with staff in school or private teachers.

Parents should make (or copy from school) a colour-coded timetable for lessons and homework. Then they can never be fobbed off with the child's contrived excuses or genuine mistakes. Ideally, the parent should be almost as aware of what is going on in the classroom as the teacher.

But I'm TELLING you. Miss Harris gave me the Nobel Prize for maths excellence. So I don't have to do maths home-work EVER again.

The ergonomics of study

Get the child into useful working habits. Make sure they have access to a proper desk or table, perhaps away from the family hubbub, if the child prefers to work in total quiet – as many dyscalculics do. Again, don't let them do homework on the floor or in bed. Provide a chair with a back support and suitable lighting. Show them how to hold a pen or pencil for optimal use, not seizing it like a spear. Remember that left-handers are likely to have a more awkward style. Look at how the paper is positioned, slightly at an angle. It is important to get the right writing grip early in life, otherwise correcting bad writing style will be hard in later life. The best style is called the tripod grip, using three fingers. The pencil should be held between the thumb and the index finger with the middle finger supporting from underneath. Rubberised pencil or pen grips can be useful; others with triangular shafts can also help.

Children can be given short and interesting passages to copy regularly – they don't have to think about the content then. In religious households *short* passages from the Bible or Koran may work well. Developing tidy and legible handwriting is an important skill, for life, but it is also useful for children to master touch typing. It is hard to undo pick-and-peck techniques in later life. Seeing words and formulae on a computer screen may help to tidy up not only their writing style but also the clarity of their thinking. Some dyspraxics may never achieve a good handwriting style, but even a reasonable fluency on the keyboard offers a handy back-up.

The star reward chart will have to be closely managed by the parent.

Partnership
Never use the word 'lazy'. Most parents will fully acknowledge how hard their children have tried, at least up to the early teens. Then the issue is often lack of motivation. For younger children a star

chart could work – with stars for neat handwriting, getting homework done on time etc. Awarded daily, it can help motivation, with X number of stickers equating to cash or a desired item. Treats without monetary value, such as an extra bedtime story, should work well with young kids. With older children the psychology will have to be more subtle, working on variations of performance-related wages – towards their allowance. Parents will have to work out their own balance between what the family income can afford and the often wild expectations of moody teenagers. The star reward chart will have to be managed closely by the parent because the average teenager will instinctively look for loopholes. Bonuses for general behavioural merits, such as cooperation, not least with siblings, or keeping their bedroom tidy, could also work.

These charts will be effective for only a time, perhaps long enough to imbue good work habits. Parents have to be equally organised as well. One of the major determinants of poor achievement in school, separate from any disabilities, is erratic and poor parenting: few books, poor diet and late bedtimes.

At a UK teachers' national conference in March 2013, a teachers' union leader talked of the 'perfect storm' of poor parenting and over-ambitious heads harming discipline in the classroom. More than two-thirds of teachers complained of 'widespread' indiscipline leading to hours of lost lesson time. Parents often failed to set an example, while many front-line teachers said they were not backed by their heads in maintaining order in class.

Good parenting is only part of the solution (especially if the school is failing). Even with well-organised parenting, it can still be challenging on the home front. Nevertheless, all the hard work will usually produce results, and it's a lot better than endless nagging and family rows. The organisation schemes may seem tedious and demand a lot of patience.Well, yes. But a balance needs to be struck as well. The whole family routine can not circle around the gravitational pull of the disability. The child is first and foremost a family member, not 'something' which comes under the label 'dyscalculic'. The partnership you have built with your child should also allow occasional humour – about poor spelling or a maths error, provided the child can comfortably join in with the joke. That will depend on the strength of your partnership and the self-confidence you have imbued in the child.

Highly successful and competent parents can also overdo turning the home into a second school. As Anthony Seldon, master of Wellington College, one of the UK's top schools, observed: 'A wise parent will accept the child is not a mini-me, that if they don't get into Oxbridge, it isn't the end of the world. What goes wrong for many is that they press their child into doing something they think they need to do to succeed rather than letting the child develop their own course through life.'

A wise parent will accept the child is not a mini-me.

Most parents probably don't do enough at home to help the dyscalculic child, and a minority may do too much. As in all things, strike a reasonable balance.

Joint exercises
Consistency is important to avoid confusing the child. Parents should work with their child's teacher to ensure they are using the same methodology. Parents can cooperate with class exercises. They should try to understand the methods used in the classroom, as well as the vocabulary, to avoid confusing their children.

For example, discussing a visit to grandparents, mentioned earlier. Or 'I spy' games:
♦ I spy a triangle, a square
♦ Something that weighs around a kilo
♦ Something longer than a metre

Using Atlas or Google Earth:

Who has to travel farthest to visit a father working away in the military or grandparent living, say in Australia, or a cousin in the USA?

Then look at populations of nearby countries; then the country in which the school is located and finally the size of the town where the school is.

Parents can involve their sons and daughters in buying a car: Adverts for fancy cars can be collected. Contrast and compare prices in newspapers, trade magazines and on line.

Or luxury houses. Children could be set a task for comparing houses in local glossy magazines. And the possible costs of internal renovation.

Then combine the work: 'What could you buy for £10,000 for a car and £750,000 for a house?'

Younger children
When parents go shopping they can get their kids to look at price tags and sales discounts. How much cheaper – by what percentage? The opening and closing times of shops – how long do we have to buy your new trainers? – could be noted.

Unusual car numbers can be spotted when you drive, as can bus and street numbers. Engine capacity of cars and motorbikes could be discussed. What are the prices of bus tickets or what is the amount deducted from electronic travel cards? When driving in the family car, talk about fuel capacity – how far could the vehicle travel on a full tank? The costs of petrol could be debated – petrol versus diesel, for example. Discuss the dashboard measurements – from temperature gauge to fuel tank. Get them to compare prices between petrol stations. You might increase their enthusiasm for the exercise if you offered them part (or all) of the savings if they spotted a discounted petrol price.

The child could assist with working out the correct tyre pressures and help with the inflation. They could be given change to put in a parking meter, giving them a perspective of costs and time.

Do they understand the sat-nav instructions, especially how many miles there are before reaching the destination? Road signs and maps can be woven into a discussion of distances. Children can compare the journey on a real map, and be asked to find their house on the correct grid. Ask them to estimate how many miles it is and how long the journey will be to their current destination.

Often, young children can be very politically correct – you can discuss speed limits and what they mean. Why are they placed outside schools? What is the speed limit on the motorway and why? You can use road signs, for example the weight or height restrictions for bridges.

Younger children may not master all the maths lessons that can be learned from sharing the driving experience, but they should at least have an idea of left and right – something that even adult dyscalculics may struggle with if they did not get the principle while young. An easy way of teaching left and right is to say: 'That's the hand you write with.' Or get the child to extend his left index finger and thumb and say, 'That shape is an L, for left.'

Adult dyscalculics will often have issues with confusion about time, especially using a 24-hour system. They will have to think hard especially about the difference between 18.00 and 16.00 and how that translates to the 12-hour watch they may be wearing.

Dyscalculic children have a skewed perspective of time so encourage the accurate estimation of time and journey lengths. This will help their maths … and punctuality.

Very young dyscalculics may not realise that hands actually move on a clock, especially if there is no second hand. Get them into talking about the big hand and the little hand and what it means. Compare 12-hour clocks with digital versions and the variations on their computers and mobile phones. (Probably, future generations will be amazed that people actually wore old-fashioned watches on their wrists – why bother when all their gadgets have digital clocks?)

One senior maths teacher said: 'Get a toy clock with moveable hands. You would not believe how successful it was using these with 15-year-olds when we were working on bus timetables.'

Well I'm telling you the hands only move when we're NOT doing maths.

Very young dyscalculics may not realise that hands actually move on a clock, ….

Talk to them about what time they will be doing something they like. Count down the time to the event. Or get them to estimate when the cake should be removed from the oven. Talk about how the time is intrinsic in the rhythm of family life. What time do we get up? What time does your father leave for work? What time do we leave for school? At what time do we usually arrive home after school? At what time will you be allowed to stay up until when you are...? (I am making an old-fashioned assumption about regular bedtimes here.)

Older children will become more involved with timing of sports events, especially if they are fans. Ask how long before the game ends? Can our side win in the ... minutes left?

When you travel, discuss the train schedule and airport display boards. They are usually excited by such details and the pleasure and fun (sometimes) of travel may encourage children to understand how time is measured.

Dyscalculics are also notoriously bad at financial planning – from an early age talk to them about managing pocket money and how phone calls are billed. Their use of smart phones probably means that they have a better understanding of phone contracts versus pay-as-you-go than parents do. So turn it around, and get your child to explain to you the advantages of various mobile phone offers. That helps their maths ... and their self-esteem, though it may not reduce compulsive phone usage.

At home deploy plenty of games and puzzles. For the younger children, happy families and snakes and ladders; for the older ones, the perennial favourite, Monopoly. Parents may assume their children are addicted to computer games but they may surprise themselves when they find that children – after initial moans –enjoy the old-fashioned competition of games such as Monopoly, and not just at Christmas time.

Count aloud as you cook: mention the timings on a microwave or the amounts shown on your weighing scales. Count aloud as you lay out screws, tiles or plant vegetables.

Watching TV – use the clocks to check on times – especially programmes the kids enjoy. Ask how long the programme lasts and how long before it starts. They can see the point in that straightaway. Or ask them how they work out the dates and time of TV programmes they plan to record. I struggled with dates; for example, I always had problems with the 1700s and 18th century, even after I had published my first history book!

As you walk – ask how many steps.

Get them into proportions: how many sweets are left? Which slice of cake is bigger?

Church or other social meetings – how many chairs are there? How many people?

In short, numbers should be seen in the context of everyday life.

Older children
Parents can use special events – for example elections, though in the US an explanation of the electoral college system may be more confusing than enlightening. In Britain talk about, say, the Monster Raving Loony Party losing its deposit in elections. The percentage required to keep a deposit could be explained. Talk about the difference between first-past-the-post and proportional representation systems.

Or, in war games, discuss the injured to killed ratios, or 3:1 ratio in offence/defence. In some computer games – risks and odds of survival could be calculated. Wi games offer lots of opportunities to use totals in comparative terms.

Challenge your older children with puzzles in the newspaper – get them started on easy Sudoku or Sujiko; bribe them with their favourite junk food (occasionally) if you have to.

Spend time on sports results – baseball and football scores or, during the Olympics, medal tables. Contrast best times in athletics. Get them into spreadsheets for scores and then compare seasons.

Hobby enthusiasts can also be seduced into spreadsheets – for example, the number of stamps from each country.

Many children, even when they are relatively young, understand electronic technology much better than their parents. But technologically competent parents could enthuse their offspring by explaining how maths and modern technology go hand in hand. Take Google for instance. Its search engine relies on several areas of advanced mathematics, such as network theory, matrix algebra and probability theory.

Become a serious shopper, a consumer king and queen. Prod the children to look at weights and measures and best buys in supermarkets. Weight-conscious daughters could be engaged in discussion of calories on food packages. Explain that cheapest is not necessarily best – discuss the potential advantages of buying in bulk. Look for deals – trying to use percentages. Or in do-it-yourself stores: 'That piece of wood is about 20 per cent too short.' Get them to dissect a receipt from a store – times, dates and costs. Discuss loyalty cards and what the points mean.

Bank balances and saving accounts: look carefully at statements. Explain credit cards (for when they are older) and the interest rates. Discuss the principles of mortgages, and buying a house (though nowadays most kids are less likely to nest until they are much older).

Presuming you give your children pocket money or allowances for teenagers, discuss their budgets. Open a bank account with them and persuade them to reconcile their accounts every month when they receive a statement in the post or download it.

Use craft activities – knitting, sewing, measuring clothes. Boys can become enthusiastic about construction kits and scale models. What do the different scales mean for aircraft, ships or model trains? What does narrow gauge mean for steam trains? Why is it that size?

Work on estimates: 'Four of us for lunch – that's about $15 per head, say– that's $60 in total. And the tip? Let's say 10 per cent.'

Frankly I'm repelled by such blatant consumerism.

Prod the child to look at weights and measures and best buys...

To inspire a wider sense of time, children could be encouraged to investigate the family tree. 'Who do you think you are?' has become a successful TV programme in the UK and US and this has enhanced the popularity of genealogy.

The possibilities of family inclusion are endless. The aim again is to demonstrate that maths is a vital, and often fun, part of life. Parents will need to know what is happening in the school curriculum and cooperate in some tests and homework. If the teachers have to show extra patience in school, it goes almost without saying that parents must display the same dogged (and creative) patience at home and on outings and shopping trips. Combine both patient strategies at home and school and the dyscalculic child will have a sporting chance of keeping up with his or her peers.

> The inability to tell the time, and to admit it, can be tough for teenagers. Bethany, who was in her final year at secondary school in Cornwall, England, was asked if other people were generally accepting of her difficulties in maths. She said: 'If I said I struggled with timetables, for example, people might be a bit more accepting of this than if I said I couldn't read the time. People tend to find this very strange, and sometimes if I slip up and say 45 past instead of quarter to, they also don't understand that.'
> *A 16-year-old in conversation with her teacher,*
> *Truro College, England.*

Achievement

The stages of success can be measured by the children's own comments:

♦ This is all Greek to me
♦ I think I can say how the teacher explained this
♦ I can explain how I understand how this works
♦ I can show you how this works – let me write it down
♦ I reckon I can do all this worksheet – probably
♦ I have really got the hang of this and can see how it can work with other problems
♦ I think I could explain how this works to the rest of my class
♦ I could even explain it to you, Mum/Dad…

Special education process

The establishing principle should be the obvious ones of forming not only a learning partnership with your child, but also with the school. Teachers are human, too. How they see you, the parent, could influence how, even if only subconsciously, they relate to your child. Try not to infer that the teacher's methods or personality are to blame. This will produce defensive behaviour by the teacher and school. For example, it would be a mistake to march into the school and announce that it has missed the dyscalculia that you have just diagnosed. Gently and politely sharing concerns is clearly the commonsense approach. Teachers in the UK feel rather under siege at the time of writing, so it is best to make a formal appointment with the class teacher or set tutor and come prepared with a list of learning difficulties you want to discuss. This may lead to (further) intervention by the SENCO (the special educational needs coordinator) or involvement with external educational psychologists. The methods of specialist assessment have been discussed in Chapter Four. Establishing a sound initial relationship makes it easier for the teacher to communicate by regular emails, for example, when private coaches raise questions with the parents who then might need to refer to the teacher.

The dyscalculic child in the UK may then be placed on the Special Needs Register (sometimes called the Inclusion Register). This policy tries to provide equal access to the National Curriculum to all children. However great the disabilities, the goal is for the child to study the same subjects as their peers and generally to participate fully in school life.

The types of intervention vary, especially in the US. In the UK, Wales, Scotland, Northern Ireland and England have separate education systems, and in even in England the systems vary in different counties. So this will be a general guide.

In Britain, the child formally diagnosed with dyscalculia may receive what is termed School Action. Parents may be a little confused by the so-called three 'waves' of special needs provision. Wave 1 is whole-class teaching that will include daily maths. Wave 2 involves 'catch-up programmes'. This aims to get the child up to the speed of his or her peers. They are daily sessions, often run by teaching assistants (with a scripted programme) and supervised by the class teacher. They last for 20-30 minutes and are small-group based (around 6-8 pupils). They can run for two to three years, but are often gauged as a booster for transition to secondary school. The third wave, School Action Plus, involves daily one-to-one teaching organised by a SENCO. This may lead to an Individual Education Plan (IEP) and referral to specialists outside the school, such as psychologists or speech therapists. The IEP will involve daily one-to-one assistance.

Parents should oversee the progress of these support strategies closely, especially the professional assessment. The parents will need to take action over any obvious barriers diagnosed, such as hearing impediments, eye disorders or perhaps issues with language if, as increasingly in parts of the UK and the US, the child's first language is not English. In addition to the individual programmes planned by the schools, parents may wish to seek extra medical advice from their doctor or specialist paediatrician.

In some cases, the schools will start a process known as statementing. This means a 'full statutory assessment leading to a statement of special educational needs'. Although it is an elaborate process taken by the school and the local authorities, it may release funds for extra individual tuition or, in extreme cases, money for attendance at special schools or units. Statements also allow extra time in exams for the students affected, which many parents see as a distinct advantage. It can also allow a 'scribe' to write answers or a 'reader' for the questions or for the student to use a laptop computer. The Access for Special Exam Arrangements has to be sorted well in advance of the exams. Statements are increasingly rare – perhaps for the bottom 1 or 2 per cent, not least because the process is expensive. Although a school in a deprived area of London, for example, may have a relatively high percentage of children – perhaps 20 per cent or more – on the special needs register, only a handful may have statements. This is just a general summary of the special needs process because UK school reforms and funding are in a constant state of flux.

The quality and financing of special needs provisions varies enormously from county to county in the UK and from one local authority to another. The current thinking (2013) is that statutory assessment and 'statements' will give way to a supposedly more holistic Education Health Care (EHC) plan. This is intended to include health and social care specialists, as well as teachers and parents/carers. This is supposed to set up one key person who can co-ordinate the layers of bureaucracy. Whether it works depends partly on available funding and how sharp-elbowed the parent is.

Parents should therefore be realistic about what the state-provided special needs teaching may achieve. 'Differentiating' the child's curriculum, if for example dyscalculia is part of a group of related conditions, may mean that the pupil will not, at least initially, be able to keep up with the average level of attainment of the rest of his or her peers.

If the dsycalculic child has to suffer all the usually well-meaning but bureaucratic school intervention processes, he or she will need a great deal of support from parents. Parents will be involved at most stages not least in filling in family questionnaires which ask about the child's health, development and behaviour as a whole. Teachers and educational psychologists will value the family knowledge of personal quirks, motivational interests and the behavioural management methods used successfully at home.

In a formal process of extra special needs assessment the dyscalculic child will be tested in the obvious skills of mental arithmetic, number skills particularly in written calculations and general mathematical reasoning. They will be checked for their spatial and visual perceptions and working memory skills. If attention disorders are also diagnosed then medical specialists may suggest medication.

The parent should ask to see a psychologist's final report. The parent should consider the summary of the test results and what the final 'diagnosis' is. This can be tough for parents. One Surrey mother, whose son had acute dyslexia, said: 'The report listed all his problems but there were no answers. I felt kicked in the stomach, sick, wounded. Having a child with learning issues is like filling a hole that keeps getting bigger.'

This distraught mother found some help via the (very expensive) American intensive courses offered by Lindamood-Bell. The British Dyslexia Association provides much more affordable advice. But generally speaking for dyslexia and dyscalculia, there is often little state provision. Despite the high incidence of special needs – perhaps 20 per cent of UK pupils – a national plan or strategy does not exist. Although all trainee teachers have to know how to tailor their teaching to suit individual needs, more special needs training is required. Otherwise, they are fully trained to cope with only 80 per cent of their charges.

Most parents will have to opt for the sometimes meagre state-provided assistance. The special individual teaching programmes will be discussed, and the parents should ask what they can do in addition to the school and local authority provisions. The hardest question for the parent to ask is what kind of short- and long-term progress – given all the intervention processes – the child can realistically expect to make. But few specialists will risk giving a concrete answer, just as doctors often shy away from offering a definitive reply to terminal patients about their possible life span. A dyscalculia diagnosis is not a death sentence, however; rather it can be the start of a very positive, long and successful life. To recall the Buddhist saying: 'What the caterpillar calls the end of the world, the Master calls a butterfly.'

Sometimes parts of the psychologist's assessment report may be very mathematical in format. And the parent may not comprehend the scales and coding, but they must not be afraid to ask detailed questions if they don't understand. To be an effective campaigner on behalf of your dyscalculic child, you need to empower yourself with detailed knowledge of what dyscalculia is and the various legal rights and legislation.

Critique of the special needs 'industry'
By 2011, 21 per cent of pupils in England (as distinct from other parts of the UK), were listed as special needs, up from 19 per cent in 2006. These include less serious problems, often designated without formal assessments outside the schools. In 100 schools the special needs rate exceeded 50 per cent. Ofsted, the official schools' inspection body, said that around half the children identified with behavioural and learning problems were actually 'no different' from other pupils. Ofsted said the causes of special needs inflation were that schools used special needs to get extra funding to improve their positions in the controversial league tables, while middle-class parents sometimes saw it as route to get free extra tuition, and

perhaps avoid the stigma of failure. The essence of Ofsted's views was that pupils were underachieving because the mainstream teaching provision was not good enough.

There are 1.7 million special needs pupils in England alone. This high number may indeed indicate some poor-quality teaching, whereby children with no real disabilities are simply falling behind. Although poor parenting is also a major factor, the quality of teachers' own education should not be ignored. In 2012, the Department of Education figures showed that more than a quarter of maths teachers – about 9,500 – did not possess a degree in maths (though some might have science qualifications). This figure for specifically unqualified maths teachers had increased by 1,000 from the previous year.

Experienced educationalists will know, however, that the most qualified teachers are not necessarily the best teachers.

The shortage of trained maths teachers is largely a case of supply and demand, not deliberate policy. Poor discipline, low status and relatively poor starting salaries discourage top maths and science graduates from entering the profession. The schools cannot attract enough scientists and mathematicians, who can earn far more in the private business sector. Hence less-qualified generalists have to 'cover' for the missing maths specialists. Some of them struggle to do so. And children suffer, especially at the top and SEN ends of the spectrum.

Francis Gilbert, who taught in a tough inner-city state school, has written a number of books attacking the state system, particularly what he calls the special needs industry. He asks, 'Is it that our children have got a lot thicker? Are teachers getting better at identifying problems? Or is some kind of chronic "SEN" inflation going on?'

Gilbert argues that the survival of so many premature babies because of medical improvements has created more special needs children in schools. Even a decade ago, he says, many such children would have died. He also believes that, especially in deprived areas, teachers are more readily identifying the correlation between deprivation and SEN. 'Poverty breeds students who really struggle to read and write,' he says. Gilbert asserts that it is a moot point whether the SEN-labelled kids have genuine difficulties or are just the victims of parents who don't value education.

At the other end of the social scale, Gilbert says that pushy middle-class parents know that a SEN diagnosis means that their children will get preferential treatment: 'extra time in exams, more attention from the teachers, and even special equipment like laptops and MP3 players'.

He also believes that teachers play the system. They get extra resources, too, and it lets them off the hook if too many kids fail their exams. It is in this context that there is so much criticism of over-diagnosis of ADHD as a politically correct euphemism for being 'completely out of control'. Or 'bouncing off the walls', as so many teachers put it … privately.

Spending on drugs to treat ADHD has increased by 65 per cent in the past four years at a cost of £31 million per annum in the UK. In the US, the use of prescription drugs to 'cure' learning difficulties has become a billion-dollar enterprise. The medicalisation of SEN implies that children's learning difficulties can be treated by drugs rather than better teaching. This trend has also prompted the explosion of alternative therapies. Gilbert attacks, wrongly in my view, 'pseudo-scientific' terms such as 'dyslexia, which means something different every time it's used'.

His bottom line is this: 'It's time we all realised no amount of jargon, drugs or massages can solve our children's problems. The only real resolution, as it always has been, is hard graft.'

Gilbert's views, perhaps best outlined in his book on how to get the best state education for your child, *Working the System* (see further reading), have some validity, but to anyone who has suffered from lifelong dyslexia or, in my case, dyscalculia, cannot doubt the severe reality of these and related conditions. Nor does a pantheon of brilliant and compassionate scientists such as Brian Butterworth have anything in common with the snake-oil salesmen who have jumped on the 'miracle-cure' bandwagon.

Alternative routes

Some parents will despair if they have to undergo the state's SEN processes, and will look in some cases for alternative therapies. If the child is a pure dyscalculic, there may not be an alternative therapy available. But if he or she has concurrent disorders, then there are many options, some of them considered bad science. The Dore method of correcting dyslexia and attention deficit disorders has been mentioned earlier, but most specialists rebut its claims to be a miracle cure. Dorothy Bishop, a leading expert on developmental disorders, wrote in 2007: 'The published studies are seriously flawed. On measures where control data are available, there is no credible evidence of significant gains in literacy associated with this intervention. There are no published studies on efficacy with the clinical groups for whom the programme is advocated. It is important that family practitioners and pediatricians are aware that the claims made for this expensive treatment are misleading.'

Monocular occlusion, however, has proved effective for a small number of dyslexic children. Children wear spectacles with one eye covered so they can see with just one eye. This is supposed to assist integration of information from both eyes and stop words jumping around and blurring. It does work, occasionally. Some poor readers also may benefit from tinted lenses. This sometimes reduces glare and distortions of seeing black print on white paper. Schools can provide worksheets on coloured paper. Other therapies relate to special diets, especially fish oils (and school meals may not always be rich in this, despite the best efforts of celebrity chefs).

Desperate parents may be tempted into all sorts of dubious therapies. The placebo effect may be useful. Yet, despite the celebrity testimonials for some of the therapies, without proper scientific support most of the alternative routes may simply be wishful thinking, and in some cases harmful or over-expensive. An alternative therapy, such as the Dore programme, which may increase children's confidence – which is good – may not improve their reading skills. You could argue that the large amount of attention lavished on a child may well make them feel more valued, happier and more confident, but the deep-rooted condition often remains the same.

Avoid miracle cures.....

The conventional routes will take time. If parents have money to spare perhaps they should avoid miracle cures and spend it on approved teaching coaches. (See the list of websites for approved sources in the appendices.) The £6 billion-a-year private-tutoring industry has long been unregulated. Some tutoring agencies do not demand proper qualifications or even Criminal Records Bureau checks. But the new Tutors Association is trying to help parents indentify the best among the estimated 1.5 million tutors in the UK. (The US has the National Tutoring Association.) Some tutors demand £150 an hour for services, despite sometimes having unsuitable or non-existent qualifications. Social networking sites such as Mumsnet and Netmums have been asked to help grade tutors to improve standards, which are being upgraded by the largest British tutoring agencies such as British Home Tutors and Enjoy Education. Well-qualified home tutoring is certainly better for dyscalculics than quack cures.

Check list for hiring tutors
♦ What precisely does your child need? Confer with the child's school teachers first. Do they need extra help to get a specific grade in a particular subject, usually maths, or do they need to learn more general skills involving comprehension, perhaps essay-writing?
♦ Get references and check them, as well as demanding an up-to-date Criminal Records Bureau certificate.
♦ Is the tutor academically qualified? Do they have a relevant degree and teaching experience? Do they know the curriculum your child is studying?
♦ They may be highly qualified but can they build a rapport with your child? Sit in on a few tutorials.
♦ Define your goals. Ask for a general teaching plan, and work out how long the tuition course may last.
♦ Use a well-established agency, preferably one which has joined the new Tutors Association. Look for experienced professional teachers not necessarily Oxbridge graduates and former public schoolboys. The latter may be enthusiastic and charming, but they will not have the experience, especially of dealing with learning disabilities.

The school and external specialists can help, but in the end it may be in the home where the best solution lies. Teachers come and go, but a parent or both parents are usually fixtures in the child's educational firmament. Some of the ideas suggested in this chapter should be about developing the whole child, not just occasionally trying to fix his/her maths problems. By working with teachers the parent should understand clearly what the main problem is, and help the child to understand it as well. Knowledge can empower and give confidence. This is so much better than giving in to bad-tempered rants and counsels of despair or alleged miracle cures.

Dyscalculic children will take up a lot of time, not least for mum's taxi service. And at home siblings may feel they are not getting their fair share of parents' quality time and attention. The dyscalculic may soon feel out-gunned by smarter younger siblings. Some parents, consciously or subconsciously, may show that they are disappointed in the lack of the dyscalculic child's school achievements, especially if the rest of the family are successful academically. There may also be financial or other worries in the home. Against this background, it will require a high standard of parenting to assure all the siblings that they are loved equally and so avoid the dyscalculic child losing his or her self-confidence. He or she will face many challenges to their self-esteem in school. The home has to be a sanctuary for their psychological recovery. Being bullied or teased at home *and* school would be a double whammy that could scar a child for life.

Parents may object that 'All this takes a lot of time.' So it does, but realising the full potential of your child will be well worth it for most parents, and for the children. Others may object to my mention of playing the card game 'happy families'. Sure, many modern families are dysfunctional – no work or too much. Also, not all families have two parents to share the burden. No family is ideal – all are more or less dysfunctional in their own ways. Most teachers comment on how broken families and poor parenting can affect children's performance. This is not a book about social breakdown, however. It is about how to help dyscalculic children from all classes and backgrounds. I came from a single-parent working-class family, and endured being a latch-key kid, and a step-child twice, as well as later trying to bring up (and love) a step-child of my own. I have no illusions about perfect families, but I am trying to describe some useful coping mechanisms for your child. Parents, it's your call.

KEY POINTS
♦ Suggested questionnaires for parents
♦ Suggestions for helping pre-school and school children at home
♦ The importance of establishing regular homework habits at home
♦ Establish partnerships with school and 'contracts' with children at home
♦ Suggestions for joint-school-home exercises
♦ How the special needs process works (and doesn't work) in the UK
♦ If parents use tutors, check them out carefully.

7: Adults with maths problems

Educational specialists know that many dyscalculic children remain dyscalculic into adulthood, but they do not know whether *all* do. Some adults learn coping and evasion mechanisms, but many, probably most, will suffer from a maths deficiency for life. Many achieve successful careers, though it is salutary to recall that 70 per cent of UK prisoners have learning disabilities. Many are functionally illiterate.

Poor maths can often mean poor pay for adults, even when they find a job. In the UK, the Institute for Fiscal Studies estimated that a decent grasp of maths at the age of ten could add more than 7 per cent to a child's eventual earning power. The 2013 report claimed that a pupil with good maths can expect to receive about £2,100 extra each year by the time they reach 30. The research also found evidence of a wage premium for schoolchildren with good reading skills, although the effect was considerably less marked than for maths. There's also the British cohort survey showing that poor maths (lowest 20 per cent) is a worse handicap than poor literacy (lowest 20 per cent).

A 2013 study of 2,000 adults by the British defence firm BAE found that despite 38 per cent having jobs that involved working with numbers, one in seven was embarrassed by poor arithmetic and 17 per cent said they regretted not taking maths more seriously at school. Almost a third of the parents questioned believed that their children's ability far exceeded their own, which was not surprising given that nationally 18 per cent of adults struggle with very basic mental addition and subtraction.

Other British surveys have even suggested that half the working population cannot do maths beyond the level taught to 11-year olds in school. That does not mean that half the population is dyscalculic; it is much less than that. There are many reasons for poor adult maths skills. People forget their school maths, especially if they don't use their skills regularly. Perhaps because of poor teaching, some just hated maths when they were young. Dyscalculics will have started failing right at the beginning of primary school. They were never able to keep up. And so the problems continued into everyday adult life.

Technology helps and hinders. Technology has radically changed the world for today's adult dyscalculics. The smartphone and the tablet are just two examples of everyday devices that enable the dyscalculic to overcome their problems with keeping time, making quick calculations when shopping, finding their way from A to B without getting lost, etc. Such devices can mask not resolve basic maths deficiencies.

Technology may not help in matters of life and death, such as when a nurse is required to measure a drug dose correctly. And it can matter, too, when a person is trying to decide on the biggest purchase in their life: a mortgage. In the current era of austerity it is not only civil servants who need to get their sums right. Poorer households may have to balance their domestic budgets, to avoid reliance on charity food banks. Everyone should be able to spot errors in bills, in garages or shops, whether deliberate or accidental. Some experts blame the education system in England which allows children to give up maths at 16.[20] Only 13 per cent of young people take A-level maths and only one in five does any maths through to age 18. In Germany the comparable figures are 90 per cent and in the US 80 per cent. While not all can be maths experts, the economy can expect employees to apply their GCSE maths acquired at 16 to practical problems.

The UK government is considering insisting on some level of maths being taught to all who study until 18. Personally, if I had been forced to continue maths at school into my sixth-form days, I would have left. Perhaps the best solution is to improve maths until the current school-leaving age of 16, not least so that basic maths can be used effectively both at work and in the home.

A large minority of the UK population is also functionally illiterate, but few will admit to not being able to read or write properly. Adult illiterates will often spend a lifetime hiding it. Ironically, if they put half the effort they use to disguise their illiteracy into actually admitting and fixing the problems, many could soon learn to read and write, especially given good family support and local adult education facilities.

Well I'm lucky really.
...At least I can
spell dyscalculia.

Illiteracy is considered shameful, but many adults – more women than men – will cheerfully admit to being bad at maths. Reading and writing are viewed as a more 'natural' and inevitable outcome of education, even for dyslexics, but a mature command of maths is not necessarily deemed to be part and parcel of growing up to become a good citizen.

And yet anyone who suffers from poor maths, especially dyscalculics, could find their lives blighted in so many ways. In Chapter Two, I listed 30 examples in a questionnaire – nearly all of which this dyscalculic author has suffered from himself.

First, dyscalculics tend to be incompetent shoppers. They prefer to pay with notes and rarely count out change in shops. They usually end up with pockets full of coins as a result. They often collect their piles of coins in a big jar and perhaps count them all at Christmas and buy themselves or family members a present.

Dyscalculic adults rarely check their supermarket receipts and tend to be slow in working out sales discounts. They are often reluctant to try out price comparison websites. Hence they usually pay too much for insurance and energy. They will rarely shop around for better deals with banks for their savings or credit cards.

In restaurants they will usually delegate the working out of bills to someone else. And if pressed to take action themselves they will typically hesitate in calculating how much tip to leave.

They have problems with ATM machines – and discard security by usually using the same PIN number for everything; as well as deploying the same passwords on computers – from Amazon to internet banking. They have real problems with checking in automatically at airports.

Dyscalculic adults will be poor at personal financial management. They will often fail to reconcile their accounts every month, and not check their credit/debit card bills in detail. They will be unlikely to fully understand the APR interest rates they may be paying. The details of various types of mortgage arrangements will elude them as well. When they go abroad they are unlikely to be proficient in foreign exchange conversions, constantly asking, 'What is this in *real* money?' And someone else would have planned the economics of the holiday.

Often dyscalculic adults make poor travellers, because they will suffer from orientation problems, never knowing where the hell they are in strange cities. They will usually cover it up with smart-arse comments: 'I am *not* lost. I just don't know where I am at the moment.' Or 'I'm just taking the scenic route'. Even at home they may not know left from right or the direction of north, south, east and west – which can cause confusion, for example, on the London Underground. One of the coping mechanisms for some dyscalculics is to develop and train their visual memories for clues, especially landmarks, to make them vaguely proficient in remembering routes.

Dyscaclulics will often carry on their early patterns of disorganisation from school days. They will lose house and car keys, forget appointments and take, literally, years to remember their own mobile phone number, which they will understandably be reluctant to change.

Even if they compensate for their poor maths skills with practical manual talents, they may well have problems with measurement, whether a correct length of wood or the size of a pane of glass.

An advantage perhaps is that dyscalculics are unlikely to become compulsive gamblers, because they will have issues with working out odds, from horses to poker to backing winners in elections. (If dyscalculics do want to risk bucking the odds, they can learn a lot from understanding gambling odds in Marcus du Sautoy's *The Number Mysteries* [Chapter 3] and Rob Eastaway and Jeremy Wyndham's *Why do Buses Come in Threes?* [Chapter 5]; see further reading.)

On the health front they will usually be unable to convert their weight and height from imperial to metric, or vice versa. They won't be able to remember their blood pressures or maybe even cholesterol levels.

They will be reluctant users of train and bus timetables and will certainly not engage in Sudoku on long journeys.

The depressing list could continue, but most dyscalculic adults will have recognised some or most of their failings by now. Most cope with resignation, humour or clever evasion. Few will take themselves off to adult education. But it can work. One senior teacher of maths at adult evening classes told me of one of her students who could speak six languages fluently but was very poor at maths. Another was a mother of young children who was classically dyscalculic – she could not tell the time. She avoided going out most days because she developed a phobia of not being back on time for her children when they returned from school. But the linguist and the clock-phobic mother learned, with determined effort, to deal with basic maths … and clocks.

Many adults may be forced to come to grips with the issue when they have to help their kids – possibly also dyscalculic – with their homework. A diligent parent will learn alongside their children, even if they pretend to know more than they really do, not least to save face. Most adult dyscalculics will be inchworms not grasshoppers, but the experience of adult life, especially in business, may help adults to see the wider picture when it comes to maths or financial matters. Experience of a lifetime of coping with dyscalculia can help smooth many rough edges.

Dyscalculia also teaches compassion for others, especially people who do suffer from serious physical or mental disabilities as well as minor educational deficiencies. I may not be a good exemplar of this, however. When I was a magazine editor, I used to fine my male colleagues a pint of beer (for me) every time they used and abused the meaning of its/it's. In the end, I had either to become an alcoholic or give up on the idea of teaching old-fashioned (but necessary) grammar to highly educated younger journalists.

A more compassionate female editor said that her dyscalculia had made her far more professional. Her lack of a number sense, she believed, made her a better editor. 'I don't assume people know what I know. I am not complacent about accuracy – I obsessively fact-check. I also write as clearly as possible because I know the obvious is not always easy to understand.'

Parents and even childless adults may find self-help maths books such as Steve Chinn's work mentioned in the further reading useful. Reading this book should help adults cope with their basic maths problems. Also, people who like reading, such as myself, but who also hate maths – and I would rather go the dentist than read a self-help maths book – could benefit from more general books about maths such as Daniel Tammet's *Thinking in Numbers: How Maths Illuminates our Lives* (see further reading). Hammet says, 'Numbers, properly considered, make us better people.' A big claim, but his book maybe justifies his comment. The dyscalculic may take pleasure in learning about societies that dispense with all maths. Apparently, the Yancos of the Amazon have for the number 3 the word *poettarrarorincoaroaae*. They do not count beyond three. Hammet also notes how the complex numerical ratios that underlie all good music are grasped intuitively by our minds, even if we can't count.

One of the most useful basic books for parents who want to teach their children maths is Carol Vorderman's *Help Your Kids With Maths*. This is a well-illustrated step-by-step manual. Much more playful is *Why do Buses Come in Threes?*, mentioned earlier. Chapter 19 is about 'magic' tricks. Parents or canny older siblings might also find David Acheson's *1089 + All That* useful, as well as Ian Stewart's *Professor Stewart's Hoard of Mathematical Treasures*.

Brian Butterworth's masterwork, *The Mathematical Brain*, is also highly informative on cultural issues without resorting to sums. Another delightful book for the dyscalculic adult who is thinking of repairing his school maths, or the parent of a dyscalculic who wants to get into maths without actually doing any sums, is Alex Bellos's *Here's Looking at Euclid* (called *Alex in Numberland* in the UK). He discusses whether numbers are innate in the human and animal world as well as admitting that 'between ten and twenty, English is a mess'. Like our non-phonetic language, English numbers are also irregular compared with many other tongues. So maybe English-speaking dyscalculics can partly blame their language! (Although you could argue that other languages are even more complex in this regard, such as French or Dutch. Whether or not they have more dyscalculics has not been proved.)

KEY POINTS
♦ Dyscalculia nearly always continues into adult life, though adults learn coping and evasion tactics
♦ Illiteracy is considered shameful, while poor maths is not a social stigma
♦ Financial weaknesses of adult dyscalculics
♦ Why adult dyscalculics often get lost
♦ But adult dyscalculics need to help their dyscalculic children.

8: Hey, you're talking about me!

(Depending on the student's age and aptitude, perhaps a parent or friend could read this deliberately short section to a severely dyslexic dyscalculic.)

S So far, the book has concentrated on teachers' and parents' responses to dyscalculia. But what about the main thrust of this book – the child or teenager coping with the disability?

If a young person is reading this – and presumably is not too distracted by dyslexia – how should you cope with this life-long condition?

First of all: *don't ever give up*. Even if you are diagnosed formally as dyscalculic, it is not the end of the world. But the disability could undermine your future – especially in further education and employment – if you don't deal with it, rather than running away from the problem.

Just as your teachers and parents are told to be determined and patient in helping you to understand maths, you must do the same. You have to practise, practise, practise, and in the end you will probably be able to catch up with the average range of your fellow pupils, especially by the crucial age of 16. Levels of success may depend, however, on whether you suffer from a mild or severe case of the condition.

You have to accept that no single magic cure exists.

Unlike dyslexia, your disability may not be a bonus in terms of achievement in arts and design, although your struggle to master some elements of maths, despite all the difficulties you were born with, should forge a stronger character for later life. Successful dyscalculics are likely to be mavericks and that can work well, whether in business or sport, for example. The coping mechanisms discussed later should make you a more rounded, tougher personality. Treat this as a gift, not a drawback. It is a difference not necessarily a disability. OK, you won't become a professor of maths, but you could become a professor in an arts subject, if you so wish, just like the chronically dyscalculic author of this book.

Your dyscalculia is not your fault. You were born with it. You can't do anything about that. It's not worth moaning about your genes. But you can do *everything* about overcoming or, in some cases, evading the disability as you grow up.

If kids bully you at school and call you dumb, you can go two ways. You can refuse to be bullied and respond early and quickly to the bully. Even if they are twice your size, they will probably back down and find easier targets. My method of responding was not physical – being fairly short and not aggressive. It was by humour, clever appeasement and making sure I was in with a popular group of other, tougher, boys. My sense of humour and sociability skills were handy tools for the rest of my life.

Or you can make a joke about your bad maths and say that, although you are not as good as your mates, you make up for it by being good at something else (sport, computers, etc). Don't let the maths deficit destroy your confidence. Being a successful maverick requires a certain cheekiness and determination. Finding unconventional alternative routes can become habit-forming and often socially useful.

So, it's 18 onto 1 is it?

Make a joke of your bad maths.

I was not bullied at school, despite regularly being ridiculed by my maths teacher. Everybody recognised that he was a bully and so I won sympathy from the rest of my maths class. I took up martial arts, but that was because of a vague interest in the sport not to face down people. Like many a dyslexic, I became a successful though not too frequent class clown, managing to win over most of my peers and even some of the teachers.

All hierarchical and competitive institutions are bound to contain some kind of bullying. Some special needs students become the bullies because they are frustrated with their lack of achievement. If their individual coaching doesn't improve their skills and self-respect, then a suitable work experience or promotion in a school sports team can often do the trick. More often, the dyscalculic is the bullied *victim*. Proper counselling (and tough remedies) from the school, and support from the parents can sometimes help the dyscalculic to feel less victimised. Trying the buddy solution – not joining a gang but finding some resolute mates and hanging around with them – can work. You can choose your own, or the school can appoint a peer mentor.

The current fashion for social network bullying can be tempered by your parents explaining that their children – perhaps you – are not at fault. And you can always switch Facebook off; you might get more homework done. This may be an oversimplification of the problem as most children are desperate to be part of their peer group activities. But Facebook and its ilk are not compulsory and may sometimes be harmful. Withdrawing from social media may make you feel even more isolated, but that may be preferable to the depression and suicidal tendencies induced by an electronic bullying that cannot be fixed.

Bullying can be psychologically devastating, especially for a dyscalculic who is already suffering from low self-confidence. As I said earlier, the more confident dyscalculic can use humour – it is rare to find that the class clown or popular wit is bullied. Schools have introduced all sorts of anti-bullying measures, working up a chain of command to the headteacher. There is often no easy solution, however, and a change of school may be the only resort.

I managed to pass my maths exam at 16 and went on to the sixth form and university. I will confess I did not then always take my own current advice – which has come from a lifetime of experience. I just could not 'get' algebra or trigonometry at all and barely coped with arithmetic, but by default (frequently writing down geometric theorems in many detentions) and an ability to 'see' shapes if not numbers, I managed to use my accidentally learned and innate understanding of shapes to secure a bare average pass in maths.

I should have tried harder in the maths I didn't get, but by age 14 I had given up and I had very little coaching or help except for occasional and sarcastic assistance from schoolmates who were good at maths.

I achieved my aim of getting O-level maths at 16 and then dropped all maths-related subjects, even economics, which interested me. In a sense I sidestepped maths after this, but as the next chapter makes clear that was not necessarily the best adult approach.

So what is the moral of this brief tale? Don't let dyscalculia beat you. You can beat it if you don't give up. Understanding that you were born with a disability should explain a lot to you, but don't let it become a label and an excuse to fail. You have it and you can deal with it, with help from others, and determination, you can reach for the stars.

After school, dyscalculia did not often impede my education and future. Often, if not always, I overcame it. And, as in judo when you use your opponent's weight against them, turn the disability to your advantage by being unconventional – when necessary. But I did not use my dyscalculia as an excuse. I rarely mentioned by maths disability. I hid it and developed coping mechanisms, the topic of the next chapter.

KEY POINTS
♦ Don't ever give up – even a severe dyscalculic can achieve a great deal
♦ You were born with it – it's not your fault. But not trying is
♦ Learn maverick habits
♦ Don't be bullied – use humour or buddy tactics to evade critics
♦ Get the best results in maths you can; you can move on to subjects you are good at and enjoy later on.

9: Let the people speak

M uch of the book so far has been about parents and teachers helping dyscalculics. In social media, especially on forums, dyscalculics around the world try to help each other.

Useful sites
https://www.facebook.com/dyscalculia

http://www.dyscalculiaforum.com/forum/viewthread.php?thread_id
=6240

Let me quote some useful extracts:

General motivational comments

Humour

'I was contemplating long and hard about posting this and all the other ones that I've had so much fun making the last few days. I worry it could be seen as ridiculing people with dyscalculia. But I decided to post it, as most of us know how good this feels. We CAN learn. This problem is not about being intelligent. We have normal and above normal IQs. If we don't, we are diagnosed with what's called acalculia, not dyscalculia – and that is a whole other issue. We are not lazy. But for some reason, how to go about a calculation just won't stick, for the majority of us. No matter how hard we try, unless we have found a coping method – like for example a memory trick, that we're unaware we found most of the time.

'No one is really 100 per cent absolutely sure why we forget, as not much research has been done. In theory there could be many reasons. Memory and lack of number sense is what research focuses on right now.

'But we know the problem is there, we know it's real, and we know the overwhelming feeling of success when something sticks. And I, for one, need to laugh a little about it sometimes.

'Humour has been my main coping method my whole life and I need it to thrive. NEED it. But as somewhat of an "advocate" for dyscalculics in this particular group, I don't want to give off the wrong impression on your behalf. The thing is, I'm just a dyscalculic myself, and these silly postings are meant for us with dyscalculia. Both to lighten up the mood and to give you representation of yourself, to finally have a place to say, "Hey, I feel this too, I am so happy I'm not alone!"'

Difficulties of living with dyscalculia

Coping

'Is dyscalculia an obstacle especially in your field? Or have you found coping methods to not let dyscalculia stop you?'

Is life worth living?

'It doesn't really matter if you're dyslexic, dyscalculic or dyspraxic – all of us feel the same depression when we're misunderstood and don't get help, and all of us can be just like any other kid if we get acceptance and help. "If you don't get the right help, it's like... not worth living, almost." '

US moan

'All of my friends in America are annoyed and confused about their local change to daylight savings time this weekend. A teeny tiny glimpse into the world of being dyscalculic.'

In India

This is what it's like being dyscalculic in India.

'There are approximately 6 million people with dyscalculia in India.

'In a country where everybody is expected to be a high-achiever in math, it comes as a surprise – and shock – when one's offspring has trouble with numbers. When scolding, threats, and extra practice fail to improve the score, parents realise that there could actually be a problem. That problem has a name – dyscalculia. Some go into denial. By Standard 9, others would've beaten the child, and pushed them; only when nothing works, do the alarm bells ring. But by then, the children would've faced ridicule, from parents, siblings, peers and the school authorities. They will feel completely unskilled in math, and their self-esteem is hugely affected.'

A teacher's lament

Heather: 'Being a Montessori teacher, and the mother of a daughter with dyscalculia, I can tell you I have tried everything I know to get her to understand numbers, time and money. She has been in Montessori school for 6 years which uses an amazing hands-on manipulative math curriculum. Her teachers (both special ed. and Montessori certified) and I have tried flash cards, Kumon and setting math to music. Nothing has worked. Just "putting down the coffee" isn't gonna do it for her.'

The world of work

Suz asked on Facebook: 'Just wondering if anybody out there is like me and struggling to find a profession within which their dyscalculia isn't an unsurpassable obstacle to their role. I'm having to leave yet another job (and there are other reasons) but a huge contributing factor is my inability to cope with numbers and the upcoming threat of ... figures landing in my lap every Monday like a sheet of German I can't understand...can anybody relate? Feels pretty lonely out at the end of my "no numbers here" branch!'

Simple questions

'How did you learn to tell time? How old were you?'

Another asked on Facebook, 'How old were you when learned to tie your shoelaces?'

(One social media response was: '30! I wish velcro had been common when I was younger.')

'Is it a human right to get help when you're dyscalculic? What kind of help? What's your own responsibility? Where do we draw the line on what's our right and what's our own responsibility?'

'How are you with recipes?'

Finding out what's wrong

Do you remember the day you first heard about dyscalculia?

There were many replies:

'Pure relief.'

'No way. I don't even remember the day I got married. Oops.'

'Years and years ago on the Oprah show. It made utter sense.'

'When the school mentioned it a couple of months ago … still in the dark with no accessible help.'

'Yep and felt relieved.'

'Last year and it was like a revelation.'

One woman discovered what it was after her daughter's teacher told her about the daughter's symptoms. She realised that she had the same condition. 'How amazing. I told my husband about it that night and I was buzzing with the knowledge that I wasn't stupid.'

A 22-year-old female undergraduate wrote: 'August last year. I had always struggled with maths. It made a lot of sense when I was diagnosed – such a weight off my shoulders to know I'm not stupid… I'm in my last year at uni and it took forever to get diagnosed.'

'An amazing educational psychologist diagnosed my son (who also has dyslexia). I realised his problems with maths were a case of can't not won't. Have to say zero support, though.'

One young student heard about it from a TV discussion. 'I was 13 and thought it could be me, but my mum thought it was some fictional disorder. Four years later, I got diagnosed with dyspraxia and I realise I probably have dyscalculia with it.'

One dyscalculic posted: 'In 1994. I was told I had "number dyslexia"; dyscalculia was unheard of at that time.'

(Another contributor asked if the previous writer meant 1941 instead of 1994.)

'I knew I had problems as a child, but I could not put my finger on it. So last year, I decided to get tested for dyscalculia. So shortly after my 44th birthday I learnt I have dyscalculia.'

Belinda said she had diagnosed herself. 'Some years ago, while tutoring a kid with dyslexia I came across the term, and symptoms … and all of sudden I had a reason why I scored all those zeros in maths, got clients to do their own change and could never master marching/dancing.'

Matt admitted. 'I was about to graduate high school in 1990, applying for college. I went thru my entire public school [ie state school in the UK] with his issue and not one single math teacher knew what my problem was. I didn't know, nobody knew… they just thought I was "special".'

Gillian wrote: 'Last year (aged 49) after diagnosis had a mix of crying and Mexican wave! Relief! I knew it was more than being thick. Note to self: Well done younger Gillian for getting through those dreadful maths classes. Fear numbers no more. You will be OK.'

Vicky wrote: 'I always had problems with math in school and was regularly embarrassed and bullied over it. I struggled through and managed to get very low grades. But none of my teachers ever picked up on any learning difficulties with me! My parents were oblivious to it too. During my high school dyslexia was the main learning difficulty and I was just deemed "weird" and "special"… none of the teachers knew about dyscalculia! The older I've got, the worse it became. Mainly during bouts of major depression. I heard of dyscalculia a year before applying for uni. I got 2 years in and had a breakdown … tutors were concerned … student support stepped in and I was finally tested for difficulties. It was a great relief to be told

I wasn't imagining my problems and that I had bad dyspraxia and dyscalculia.'

'I was 28 when officially diagnosed. I'm now 31 and proud that I have an amazing daughter, who at 13 is very intelligent and highly gifted in maths. She has to help ME do her maths homework!'

Colleen posted: 'Last year, after I talked to the doctor. I got good grades in school – as long as I stuck to English, history (details, not dates) and avoided math and some sciences.'

A BBC report helped Michael. 'What struck home though were the other symptoms also associated to this, like clock-face issues, disassociation of self to maps.'

Another dyscalculic remembered the discovery vividly: 'I found the term by accident while looking up something else and began researching it. I clicked excitedly through articles and testimonials, all the while thinking "it's not my fault. IT'S NOT MY FAULT!"'

Jacqui remembered the day and the feeling of relief. 'I knew it was something real and I wasn't dumb or stupid.'

Jean said: 'I was 57 years old and I cried with relief. Now want to do all I can to raise awareness of its existence and get support for all children who struggle with this debilitating condition.'

'I was 17 and just starting college,' Jennifer said. 'I guessed well enough on my placement test to land myself in a statistics class ... panicked and went to my professor who just happened to have two PhDs in learning disorders. She diagnosed me and switched my class. I was angry no one found it sooner but also tremendously relieved that it was NOT me being stupid. My 6th grade teacher actually told my mom I was retarded and unteachable and should be in a special school. My daughters all have this in one form or another and getting help today is no easier than it was in the 70s for me.'

'Seventh grade and my mom told me I fit into a lot of the symptoms. As did she. I love genetics.' That was Marilyn.

Liza was triumphant when she found out: 'I think I was 47 or 48 in 2007 when I did a search on it on the Internet and I discovered for the first time in my 48 years I was not the stupid person my mother and ex-husband made me out to be. What a sense of relief to know they were wrong about me!!!'

What's the one thing you wish you could have said to your old maths teachers?

'Please believe me when I say I am trying as hard as I can. There is nothing in the world I want more than to understand this. I know I look and act indifferent, but this is the only way I know how to not break out in tears. I want your approval. I don't want this special attention. I don't want to be a burden. I want to understand.'

'I'm not thick, stupid, lazy or disruptive,' Karen said. 'I am confused, scared and desperately want to understand like all my friends do. I want to be able to do my homework without crying and I want help.'

Donna would have told her teachers: 'When I shouted at you "I can't do it" I needed your help, support and understanding not your disapproval and punishment. And to another ... It really wasn't helpful to sit me at the back of the class because YOU didn't know how to help me. Perhaps both of you should have realised I had a real problem!!'

Julie replied: 'Hope karma comes back and bites you in the bum for ridiculing me and making me feel so small.'

Tatiana said simply: 'I really wasn't being lazy.'

Mary was also very direct: 'I would have said, "Stop speaking in tongues." '

Gillian wrote: 'Now do you believe me?'

Teresa said: 'Do you see I wasn't lying or kidding when I said I was working as hard as I could. Do YOU get it?'

Mindy wrote: 'While my first instinct is FU! Particularly for the teacher who told me: "This is dummy math. Any dummy can do it." But...I really should educate this ignorant man on dyscalculia and show him some materials on it and how to identify it in students and techniques. I recently found out he is still teaching at the school I went to and I graduated 14 years ago. So tempted to write an email!'

A parent's compassion

Sarah said to her daughter when she complained about her dyscalculia: 'I was the same...crying in school, getting "sick" before math class so my mom would come get me, etc. You are incredibly bright and intelligent and just plain awesome. We will get through this and become better people because of it. Your Pop and I desire and love your compassion for others WAY more than your aptitude for math concepts ... let's keep our eyes focused on the most important things, okay?'

Success stories

Social media often comment on the inspiration of famous dyscalculics. The singer Cher is often mentioned. On Facebook, one reader encouraged her friends to read Cher's autobiography. But Bill replied: 'A dyslexic dyscalculic trying to read an auto-biography written by a dyscalculic dyslexic … I don't think I'm going to make it through that piece of literature. Lol.'

Brenda: 'It's a funny thing ... occasionally when a step by step mathematical calculation is explained to me the light goes on briefly, I understand momentarily and then it all just disappears like it never happened! I don't get it, but I've stopped stressing about it!'

KEY POINTS

♦ Social media support networks can be very helpful and empowering
♦ Not least to persuade dyscalculics they are not dumb, or alone
♦ Teachers should take on board some of the comments and maybe rethink their approach to 'special needs' pupils with dyscalculia

10: Coping mechanisms

Success stories

We are told that some great men and women had difficulties in their early school lives – facing problems with not only calculating, but also reading, writing, spelling, paying attention and using memory. Artistic and scientific geniuses such as Leonardo da Vinci often displayed academic disabilities. Leonardo's mirror/reverse writing and poor spelling suggest he was dyslexic. Michael Faraday, another scientific genius, had an unpromising academic start in life. As did Albert Einstein. He made the famous comment: 'If a cluttered desk is a sign of a cluttered mind, what then is an empty desk a sign of?' Some of the maths deficiencies of the young Albert have been exaggerated, but he was on to something about untidy desks. Nearly all dyscalculics tend to be disorganised in their personal lives and in their study habits. I must confess that I am writing this on a very cluttered desk.

Internet chatter claims that Bill Gates is a dyscalculic, but there is no evidence – except gossip – to support this and Bill is a hard man to pin down for an interview so I can't check with him. There is more evidence to suggest that the Danish writer Hans Christian Anderson had problems with numbers, but not words. Benjamin Franklin also had a reputation for being dumb at school maths, which didn't prevent him from becoming a scientific genius, polymath and founding father of the US. It must also be noted that lack of money forced him to leave school at ten, so he was largely self-taught.

And modern examples? Well, it is well-known that international businessman Sir Richard Branson overcame his chronic dyslexia to become a billionaire. It is, however, harder to find dyscalculic billionares. Brian Cox is considered the 'coolest prof on the planet' because of his popular British TV programmes on science. But he has both a dodgy hair style and dodgy A-level results in maths. Cox admitted that he had to swot hard at maths, to get below average results (a D). 'Very few people are instinctively good at maths,' he said. To take an American example: Tim Doner is a New York City schoolboy who has learned over 20 languages and became a YouTube hit. A linguistic genius obviously but he admitted, 'I wish my maths was as good as my Swahili.'

If dyscalculics need inspiration or encouragement when they worry about understanding money they might listen on YouTube to Mick Hucknall, the lead singer of Simply Red, performing 'Money's Too Tight (To Mention)' at the Sydney Opera House. Hucknall has talked openly about his dyscalculia; and the Opera House designer, Jørn Utzon, was said to be very poor at maths.

Dyscalculics don't have to be pop singers or architects to make a mark. Louis Barnett left school at 11 because he was bullied because of his learning problems (dyscalculia, dyslexia and dyspraxia). He was then home-schooled in Staffordshire, England; in his spare time he began making cakes for his family and friends. As with some autistics, an obsession became a successful career. Barnett fell in love with the art of chocolate making. In 2005, aged 12, he launched his first company, Chocolit (a nod to his dyslexia). Soon he was well-known, supplying high-class stores such as Selfridges, as well as employing his family in his work force. He is now 21 and treated like a rock star when he travels to Mexico to meet his chocolate suppliers. He has established a world brand, from a hobby which was 'something to take me away from the tough time I was having'.[21]

A younger self-starter is ten-year-old Joseph Macintosh. At five the English schoolboy was so disruptive that he was considered almost unteachable. He had a range of what he calls 'dyses', which sound teaching, intensive occupational therapy and medication helped him to overcome. His school record improved rapidly as did his maths because he set up his own business plan (with a little help from his father) to launch a registered charity called Beat Dys – to help children with learning difficulties, including dyscalculia and dyslexia.[22]

Overcoming dyscalculia
If dyscalculia is an inherited brain wiring problem, then – short of revolutionary brain surgery or wonder drugs invented in the future – sufferers have to learn to cope, not just at school but throughout their lives.

The best coping mechanism is early diagnosis and suitable individual intervention, perhaps long-term. The child has to understand that they are likely to be born dyscalculic. It isn't their fault. But not dealing with it is. Armed with a personal teaching plan, from nursery through to primary, the child should not be allowed to fall below the level of his peers, if possible. They should not miss out on the foundation concepts of maths. The confidence of the child must be maintained, otherwise a vicious circle of anxiety and failure will result. The lack of motivation and self-confidence can undermine a whole life.

> **'Maths is a subject that is quite sequential – you have to learn how to do step one before you can do step two and so on. And as soon as you don't understand one step, you fall behind. Too many youngsters just don't have the confidence to put their hands up and say, "I don't get it". They think the problem is with them.'**
> *Sir Terry Leahy, the former chief executive of Tesco*

The earlier chapters of his book have gone into some detail about how teachers can alleviate and often resolve many of the problems. Patient one-to-one teaching is clearly best practice. I also discussed the special-needs process as an addition to standard teaching.

The good parent will be a key part of coping for the dyscalculic child. If he or she suffers taunts at school, the home has to be a haven from such assaults on their self-esteem. The parent must explain to the child – and his or her siblings – that dyscalculia is not the child's fault. All the children in the home should understand that it is probably a part of family history. 'Uncle Fred had dyscalculia, too, and he is now a top businessman.' 'Mum is poor at maths but she has been successful in her career.' Other families have the same problem, too, the parent can explain. And they could go on to mention successful celebrities who have overcome similar or the same conditions.

The siblings could be asked to help if they are good at maths. That will make them realise how tough it is for dyscalculics to understand the world of maths. It will also inform the family how hard the dyscalculic child may be working. The parent should reinforce this by closely monitoring the school work and praise areas of improvement in maths and probably greater achievements in other subjects. Obviously some sensitivity is required in discussing tests where the dyscalculic comes in the bottom five per cent.

The parent must try hard to keep on top of the maths curriculum, homework and class results. That stops the child feeling isolated and so be tempted to give up. Parents should not take all the stress. That is why it is useful to share the problems and joint homework with siblings and perhaps grandparents. Some organisations (see appendices) host local support groups for parents who have children with learning difficulties. They may prove useful if only to share experiences (and frustrations) without undermining family solidarity.

As pure dyscalculia is relatively rare, a child may also be challenged in other areas besides maths, including poor reading. If so, then try to make a short joint reading session a daily habit, not necessarily at bedtime. You could curl up in a comfortable part of the house although not at the time of the child's favourite TV programme. Once you overcome, with younger children, the obsessions with princesses, dinosaurs and robots, the object should be reading for pleasure. You could try to form a family reading group – now that does sound ambitious – with teenage children, providing that they are not too keen on Twilight genres. Harry Potter can be overdone, although the British royal dyslexic, Princess Beatrice, said she fell in love with reading in general after being seduced by J K Rowling's wizard. The point is to make literacy and numeracy part of the family, part of life – and not just isolated to one child's problems with maths in school. Literacy will obviously extend beyond books to interesting menus, recipes and lively and informative blogs, or even well-honed tweets.

It may be useful to share novels where the central characters are youngsters, with dyscalculia, who overcame their conditions. Two recent examples are *Girl Wonder*, by Alexa Martin, and Kathryn Erskine's *The Absolute Value of Mike*. Another example is the sometimes harrowing memoir, *My Thirteenth Winter*, discussed earlier.

It may also be useful for the dyscalculic to keep a diary of his or her experiences with the disability, not because of possible publication later, as in the case of Samantha Abeel's *My Thirteen Winter*, but because it may have a major therapeutic value in itself, according to recent research.[23]

The young dyscalculic probably won't want to read publicly from his diary but parents might get the family to review books and to discuss the reviews and maybe even the whole books. Literate parents could even push their luck and get the family to discuss alternative endings to the books they are reading, including some of the classics. The 'what ifs' can provide lively debate and develop reasoning skills. This can be extended to other subjects besides English. Take history: what if Hitler had won the war and conquered Britain? There are a whole range of counter-histories in novel and 'historical' forms that could excite interest. In geography, what if France and Britain had remained physically joined together? The debates on global warming could take in a range of school subjects.

This may all be more than a touch idealistic for most busy families. But the ideal of an educationally striving and well-rounded family will help a child with learning difficulties. A successful domestic environment can compensate somewhat (or even entirely) for a poor school environment. It is best if both are successful and complement each other.

The role of the home should be that of a support system. Parents need to follow the following common-sense rules:
♦ Discuss the issues as openly and honestly as possible
♦ View special needs provisions not as a stigma but as a vital tool
♦ Stimulate children as much as possible
♦ Be consistent in routines, rules and attitudes
♦ Be positive and open-minded about the child's (and their own) learning
♦ Aim high at home and in cooperation with the school. In some cases the dyscalculic child may even need to join classes for the gifted and talented.

If parents get all these rules right, paradoxically, it may not even matter where the child goes to school.

In the final analysis, your child should know that you believe in him or her. Only then can they believe in themselves.

The chapter written for the dyscalculic child – 'Hey, You're Talking About Me!' – should be read by a dyscalculic older child, or read to him if he is also severely dyslexic. No amount of worthy intervention by school, parents and educational psychologists will help if the child cannot be persuaded to help himself. Some don't and become unsuccessful adults, especially those who end up on the wrong side of dole queues. Or the law. As Brian Butterworth noted, 'One of the first dyscalculics we saw, many years ago, was in prison for shoplifting. It turned out that he was too embarrassed to go the till because of his problems with money.'

Helping yourself

Recent research in behavioural genetics, characterised by the work of an eminent American, Professor Robert Plomin, suggests that heritability of factors such as IQ and even weight goes up lineally across the lifespan by perhaps up to 80 per cent. That could imply that all the best private coaching, expensive schools and tiger mothering may help in the early years, but in the end genes will out. As you grow up the genetic differences, almost regardless of earlier domestic or school environments, become bigger as you create your own environment which correlates with your genotype. As Professor Plomin summarised his thesis: 'Bright kids read more, then hang out with kids who read more.' Plomin, who has worked intensively on studies of twins in the UK, has shocked the educational establishments in both Britain and America, not least with his futuristic talk of DNA sequencing of all newborns to define disabilities as early as possible.[24]

Some of Plomin's conclusions may indeed appear disconcerting, but his emphasis on heritability at least debunks many myths about disability, such as that bad parenting causes autism. Parents once had to deal with multiple problems of coping with the chaotic lifestyle of a child with severe autism as well as the world thinking it was their fault. And ADD us still attributed to bad parenting, poor teaching or too many fizzy drinks and chocolates. ADD is heritable, just like dyscalculia.

Dyscalculia is inherited and in the genes. Just like IQ. But no one thinks IQ is a valid means of measuring a person's overall worth. Not being good at maths is just one small part of the personality. To adapt a saying, you could even assert that maths contains much that neither hurts one if one does not know it, nor helps one if one does know it. You can also argue that grit and determination may well be inherited qualities as well. So, despite the recent research in genetics, I would suggest that family is not inevitably destiny.

I have outlined many examples where the family and the state can help dyscalculics to transcend their inherited disability. But let me return to the prime lesson of my own experience of dyscalculia – the determination to help myself. Self-help is often derided as an old-fashioned, even Victorian, value. It has even become fashionable to argue that individuals are incapable of using free will to improve their character and determine their own destinies. I repeat: there is little doubt that developmental dyscalculia is genetically inspired. But that does not mean a surrender to what is called 'neuro-politics' – that everything from a person's sexuality to politics is genetically determined, the gift, or curse, of cranial wiring rather than something a person can command or cultivate. Some modern theories, for example on addiction, allege that humans are not assertive creatures but more like an amoeba in test-tube shaped by all sorts of forces beyond their control. And, although I have described the importance of early diagnosis and intervention, I do not entirely accept the 'early-years theory'. Plomin's research has indicated that early years perhaps matter less compared with a lifespan of inherited factors. Nevertheless, the first five years may be important, but not – as the Jesuits famously claimed – the inevitable determinant of the future. People with any of the 'dyses' should not surrender to fate, to the idea that we are entirely sealed by genetics and past experience rather than an ability to shape our lives by thought and action. In short, we have some choice whether to grow out of, or into, our dyscalculia problems, even though the condition is generally permanent.

This is not the place to say more about the nature/nature debate but just to emphasise the importance of confident shaping in a young person's life. This book has been about securing that confidence by his or her own means, and by parents and teachers aiding that confidence-building. Ultimately, an individual has to forge himself into a fully rounded personality.

And how many of the adults caring for the dyscalculic child are fully rounded? Some dyscalculic youngsters may receive precious little help from school or home. Even caring parents will be reluctant to take up adult education if they are incompetent in maths. But as they have possibly passed on the dyscalculic genes to their children, then online education, or even number games, and working with their children's homework, might avoid the creation of a vicious circle. What an achievement if the mathematically challenged mother or father overcomes the problem in adulthood, both for their own sakes and for the education of, and example for, their offspring.

In this book I have tried to distil a lifetime of coping with acute dyscalculia to help children sidestep a dysfunctional career and lifestyle pattern. I didn't have much extra help from school or home, but I was determined that I would succeed despite my chronic maths problems. The practical information suggested in this book can help dyscalculic youngsters avoid having to take that lonely path themselves.

KEY POINTS
♦ The best coping mechanism is a personalised one-to-one teaching programme when the dyscalculic child is young
♦ Support from school should be complemented by confidence-building measures at home
♦ Only if the family believes in the child can he or she learn to believe in themselves
♦ Despite the inheritance factors, self-confidence can inspire self-help
♦ Parents should improve their own maths not least to help their children
♦ In the end, however, the child has to learn to help himself.

CONCLUSION : Making it all add up

Children with learning disabilities I have detailed can be very creative, imaginative and insightful.[25] Teachers and parents should share in the intervention programmes with the children. Dyscalculic children must feel they have a stake, not that the system is just making them jump through endless and pointless hoops.

Even if the school and family provide the very best support, the dyscalculic will have to make his or her own decisions as they get older. They can either cave into the pressures or overcome them. As dyscalculics will probably be of at least average intelligence, they should be able to work out a range of coping mechanisms to beat, manage or evade the challenges they face. Many dyscalculics learn to be mavericks. Dyslexics may be 'gifted' with extra artistic skills, but I would argue – from much personal experience – that the maverick dyscalculic can forge a successful career path. They will never become maths professors, but they can – if they are determined enough – pretty much do anything else. The support mechanisms outlined in this book should teach the dyscalculic to cope with average maths challenges, but they should also inculcate a self-reliance that may be more useful than any amount of number facts. Above all, he or she should believe, early on, that just because they can't count, it doesn't mean they don't count. He or she should believe they should count for anything and everything they want.

Finally, it *will* all add up.

APPENDICES

Appendix 1: Useful websites

General information

Brian Butterworth's research
www.mathematicalbrain.com/

Dynamo maths
http://www.dynamomaths.co.uk/dyscalculia-def.html

Emerson House
http://www.emersonhouse.co.uk/

Learning works
http://www.learning-works.org.uk/

Forums for teachers and parents
http://www.dystalk.com/topics/5-dyscalculia

British Dyslexia Association
http://www.bdadyslexia.org.uk/

The Dyscalculia Forum
http://www.dyscalculiaforum.com/news.php

Dyslexia Action
http://dyslexiaaction.org.uk/

The Helen Arkell Dyslexia Centre
http://www.bdadyslexia.org.uk/

http://www.dyscalculia.org/
This has lots of general information

http://www.mathsproject.com/

dysTalk
http://www.dystalk.com/topics/5-dyscalculia

Math and Movement
http://mathandmovement.com/whatis.html

Government sources

http://www.education.gov.uk/lamb/module4/M04U16.html

Teaching aids for parents

Dynamo Maths
http://www.dynamomaths.co.uk/

The Number Race
http://www.thenumberrace.com/nr/home.php

http://number-sense.co.uk/

Forum for dyscalculics on Facebook

http://www.facebook.com/dyscalculia

Private coaching

www.kumon.co.uk/

www.bht.co.uk/

www.enjoyeducation.co.uk/

Appendix 2: Teaching tips

Answering the question

Even as an examiner at post-graduate level, I find it is common that students do not read the question properly and so fly off on in all sorts of strange directions. Preparing a dyscalculic child – at the start of his/her learning career – is important. Typically a dyscalculic student will see what he *wants* to see – that is, something he thinks he can do. If the child is also dyslexic then the chances of misunderstanding or misreading the question are obviously enhanced. So here is a basic lesson for solving maths problems in primary school and secondary:

1 Read the problem slowly and carefully, and then re-read it. Read it aloud to yourself, quietly, once again. Are you sure you know what you are being asked to do?
2 <u>Underline</u> important words
3 Read each word individually to make sure you know what they mean
4 <u>Underline</u> or highlight the key numbers
5 Break down a possible solution into separate steps
6 If in class and you are not sure how to solve the problem, discuss with your classmate or ask the teacher (once you have really tried yourself)
7 Make sure you know what the problem is really about – perhaps division
8 Decide on method and correct symbol ($\div$)
9 Do a quick estimate – in round numbers – to work out a rough answer
10 Neatly and carefully work out the answer, using diagrams etc if a geometric question
11 If a long division question, check whether your initial estimate roughly matches your final answer
12 Go back to the original question and confirm whether you have answered what you were supposed to.

The above may seem obvious, but many questions are mis-understood even at the best universities. Learning how to use a belt-

and-braces approach early on can set up a pupil for life, and this is especially important for the dyscalculic child.

Maths Short Cuts

An odd plus an even number always has an odd number answer

'Sign Language'

One common way to explain the four main symbols:

+ and x consist of two lines and you can do the calculation in two ways to produce the same answer: 3 + 4 or 4 + 3 will produce the same answer, or 3 x 4 and 4 x 3 ditto.

Whereas - and ÷ have only one line and you do can the calculation in only ONE way:

7-4 but not 4-7; 12 divided by 3 but not 3 divided by 12.

This helps to establish a basic and useful fact and which may allow the pupil to start seeing maths patterns.

Dear Algebra,
Please stop asking us to find your X.
She's never coming back and don't ask Y.

Appendix 3

50+ Symptoms of Dyscalculia

The most common symptoms in approximate order of occurrence/ significance.

The following list is not definitive or comprehensive, but it should provide a handy general guide.

Basic maths
❏ You cannot guess amounts quickly (without counting) for even small quantities
❏ It takes an abnormally long time to tell which number is larger or smaller
❏ You have trouble rounding even fairly low numbers
❏ You rely on strategies such as finger counting
❏ You are confused about mathematical symbols: +,-,÷ and x
❏ Therefore, you have difficulty with addition, subtraction etc
❏ Sometimes you transpose/reverse numbers – writing 63 for 36, for example
❏ Sometimes you confuse similar numbers, for example 3 and 8
❏ You may have difficulty in conceptualising basic formulae
❏ You have problems with remembering maths operations – mastering it one day and completely forgetting it the next
❏ You can't remember times tables
❏ You have difficulty using calculators. Dyscalculics have to check a number of times – until the answers are the same two or three times
❏ You may have difficulties in reading an ordinary clock, and problems with 24-hour version even as an adult
❏ You are slow in counting backwards from ten down
❏ You cannot easily keep score during games. Football is easier, but cricket and tennis are more complex. You may have a good grasp of the principles of the games, but find it hard to plan ahead more than a few moves, in chess, for example

❏ You suffer from maths phobia/anxiety; chronic avoidance of maths homework

❏ You find it hard to copy down on paper a list of numbers when read out by the teacher, or even from the board

❏ Sometimes you see a number written down, but when you copy it on paper the numbers may be in the wrong order

❏ You are anxious about using the phone sometimes – you write down transposed numbers. You need to check and re-check the numbers.

❏ You take much longer than other students when asked to do basic mental arithmetic

❏ You don't understand what fractions, odd and even numbers, let alone square roots, mean

❏ When doing a maths problem in school, even ones you have done before, you forget where you have got to, and then you have trouble finishing off the problem

❏ When a problem is completed – and even when it is correct – your working is very untidy and then you are still not sure how you worked out the answer

❏ Sometimes when the answer is right you can't explain how it was done

❏ Big numbers confuse you

❏ You find percentages are very difficult to work out (though many non-dyscalculic adults may also admit to this)

❏ When the questions starts, 'If a man with a tractor can plough two fields in two days, then how long …' you switch off immediately.

Visual-spatial issues

❏ Many children with dyscalculia are not good at puzzles

❏ Conceptualising time and the passage of time when related to numbers is difficult

❏ Dyscalculics will often be chronically unpunctual, especially when they are younger

❏ They may struggle to visualise geographical locations – from states/counties in their own country to lay-out of streets in their own neighbourhood. In short, they often get lost, even in school
❏ Reading a map (especially with grid references) is challenging
❏ Can't discern left from right.

The basic maths issues can be tested fairly accurately for dyscalculia symptoms. It is often harder to define visual-spatial problems related to dyscalculia alone; they may relate to concurrent conditions. The following adult behavioural list is scientifically much more arguable, but many of the problems – if grouped together – may indicate dyscalculia in adults.

Adult behaviour

❏ Inability to remember numbers, even a home phone number or car registration number
❏ Habitually offering bank notes in a shop rather than working out change for a small purchase
❏ Avoiding reconciling accounts and personal finances at the end of every month
❏ Using same PIN number for all cards
❏ Difficulties with working out north/south/east/west or even left and right
❏ Always delegating somebody else to work out bills in a restaurant
❏ Phobia about maths, especially mental arithmetic
❏ Maths deficiencies probably run in the family
❏ Problems in working out recipes using detailed weights and measures although women tend to do better by learning by rote
❏ Difficulties in working out DIY details – for example, how large a pane of glass must be
❏ Problems converting weight from imperial to metric (stones to kilos)
❏ Don't fully understand how mortgages work, especially the advantages/disadvantages of tracker mortgages
❏ Tendency to avoid using bus/train timetables

❏ Not proficient in working out exchange rates for foreign currencies

❏ Don't usually check your supermarket shopping receipts

❏ Slow in working out sales discounts in shops.

❏ Can't remember when you are told what your blood pressure or cholesterol levels are

❏ Trouble working out change in a shop, especially if other people are waiting/watching.

Some adults may 'fess up and say, 'Yeah, fair cop, I can't do most of the above list, so I must suffer from genuine maths problems, maybe even dyscalculia.' Others – perhaps with dyscalculia – may say, 'I could do all the above, if I had time or could be bothered.'

Being good at maths

By way of contrast, what shows you are good at maths? Some examples:

❏ An ability to operate with numbers and other symbols

❏ Ability for sequential, segmented and logical reasoning

❏ An ability to shorten or reverse the reasoning process

❏ Possess a mathematical memory

❏ Flexibility with spatial and abstract concepts.

Relationships are a lot like Algebra.

Have you ever looked at your X and wondered Y?

Glossary of terms and abbreviations

Abbreviations and acronyms

ADD   Attention Deficit Disorder

APR   Annual Percentage Rate

DD   Developmental Dyscalculia

G & T Gifted and Talented

IEP   Individual Education Plan

MLD   Mathematics (or Moderate) Learning Difficulties

SEN   Special educational needs

SENCO Special Educational Needs Coordinator

VAT   Value Added Tax

Glossary of terms

Not all the terms are necessarily used in this book. I have added some that confused me as a child and adult.

Algebra
The branch of maths that deals with generalised arithmetic by deploying letters or symbols to represent numbers.

Bridging
A teaching method based on a linear progression of numbers in which subtraction or addition is done not in single number steps but in convenient jumps, using, say, 10 as the stepping stone.

Chunking
A term often used in teaching to describe grouping together numbers, as opposed to letting the pupil count in ones.

Clinometer
A hand-held instrument, usually of plastic, which measures the angle of elevation of some point, in relation to the horizontal plane of the observer. I always was confused by horizontal and perpendicular but now know they are the flat plane and an upright one.

Co-morbidity
A condition that occurs at the same time as another condition, but may not be directly related (for example, dyslexia and dyscalculia).

Compasses
Shorthand for a pair of compasses, an instrument with two legs, hinged together at one end, with the two other ends made up of a sharp point and the other containing a drawing tool, usually a pencil.

Composition
Addition (as opposed to decomposition: subtraction).

Computer numbers
A **bit** is the smallest piece of information that can be stored or transmitted. Usually depicted as 1 or 0

Byte is the number of bits grouped together to make a single unit of data. Usually 1 byte = 8 bits

Kb = Kilobyte which is 1024 bytes or often just 1000 bytes

Mb = Megabyte which is 1048576 bytes or often 1 million

Gb = Gigabyte which is 1073741824 bytes or a 1000 million

Counting all
When a pupil starts counting in ones from the first number in a calculation. It often indicates a poor number sense.

Decimal point
A dot or full stop used to show that the values which follow make up the decimal fraction (a comma is used in other systems such as metric).

Disc/disk
Computer disk or a compact disc.

Estimation
An approximation of a quantity that has been decided by judgement, rather than by a process to work out an accurate answer.

Euclid
A Greek mathematician (born c. 365 BC) who founded a school of mathematics at Alexandria. He wrote the longest-lasting textbook on maths.

Gross
The amount of weight or money that is the total before any deductions are made. As opposed to net which is the amount after deductions are made.

Formula
A statement, usually written as an equation, that gives the exact relationship between certain quantities, so that when one or more variables are known, the value of an unknown quantity can be discovered.

Fraction
A measure of how something is to be divided up or shared out.

Geometry
The branch of maths that studies the properties and relationships of points, lines and surfaces in space.

Inequality symbols
< is less than; > is greater than.

Lune
The crescent shape formed when two circles overlap. Not of much use except for a (tough) pub quiz.

Magic square
A set of numbers arranged in the form of a square so that the total of every row, column and diagonal is the same.

Maths anxiety
A highly stressed fear of maths that often creates a psychological barrier that can make it very difficult for the child or adult to solve mathematical problems.

Memory
Long-term
The ability to retain knowledge that can be retrieved again over a long period of time.
Short-term
The ability to recall facts that are needed temporarily for immediate use, for example a telephone number or a car registration number.
Working
The retention required to carry out step-by-step procedures and to reason out solutions to problems.

Mind map
A diagram used to visually outline information. It is often created around a single word or text, placed in the centre, to which associated ideas, words and concepts are added. Major categories radiate from a central node, and lesser categories are sub-branches of larger branches.

Mnemonic
A device that is intended to aid a person's memory.

30 days hath September
April, June and November
All the rest have 31

Excepting February which,
Has 28 days clear,
But 29 in each leap year

Multiple
A number made by multiplying together two other numbers.

Number
For a dyscalculic, the comparative of numb.

Number line
Numbers are marked on a line (paper or plastic for example) at regular intervals. Fractions can also be shown in intervals between the whole numbers.

Overlearning
Is a concept that subscribes to the belief that one should practise newly acquired skills well beyond the point of initial mastery, leading to automaticity.

Per cent
A special type of fraction that measures the number of parts in every hundred that is to be used.

Processing speed
The pace that children can absorb written, visual, or oral information. If the speed is too fast then children may fail to understand or remember the data provided.

Plurals
axis	axes
datum	data (though more commonly now, 'data is')
die	dice
focus	foci
formula	formulae
hypothesis	hypotheses
locus	loci
radius	radii or radiuses

Proof
A sequence of arguments made up of axioms and assumptions leading to the verifiable truth of one final statement.

Progression
Arithmetic progression (AP) is a sequence where each new number after the first is made by adding on the same amount to the previous group (3,7,11,15, 19 where the constant is four). Geometric progression (GP) is where you multiply by rather than add a constant number.

Protractor
A semi-circle usually, of clear plastic, with marks for measuring angles.

Random Variations
Variations in any series that happen in an uncontrollable and unpredictable way throughout. Besides maths teachers, could also be used by government budget forecasters.

Rounding (up)
Shortening or simplifying a number, for example making 89.5 the round figure of 90.

Square
A four sided shape whose edges are all the same length and all angles are 90 dgrees.

Square numbers
Numbers that can be represented by the correct amount of dots laid out in rows and columns to make a square. Generated by multiplying a number by itself. For example, 4 x 4 = 16. 16 is a square number.

Subitization
The ability to recognise a quantity without counting.

Subvocalizing
Talking to yourself quietly while working out problems.

Value Added Tax
A tax paid on goods or services bought by a customer, which then is (or should be) paid by the supplier of the goods or services to the government.

Visual and spatial awareness
The ability to see and relate to the distances between objects.

Whole number
A number that has no fraction or decimal point attached. Sometimes called an integer.

Further reading

For teachers
Ronit Bird, *Overcoming Difficulties with Number: Supporting Dyscalculia and Students who Struggle with Maths* (Sage, London, 2011). A very practical (and individualistic) primer for teachers, especially new ones.

Brian Butterworth and Dorian Yeo, *Dyscalculia Guidance: Helping pupils with specific learning difficulties in maths* (GL Assessment, London, 2004)

Steve Chinn, 'Mathematics learning difficulties and dyscalculia' in Lindsay Peer and Gavin Reid, *Special Educational Needs: Guide for Inclusive Practice* (Sage, London 2012)

Jane Emerson and Patricia Babtie, *The Dyscalculia Assessment* (Continuum, London, 2010). A very useful handbook for assessment at primary level.

Glynnis Hannell, *Dyscalculia: Action Plans for Successful Learning in Mathematics* (Routledge, Abingdon, 2013)

Anne Henderson, *Dyslexia, Dyscalculia and Mathematics: A Practical Guide* (Second Edition) (Routledge, Abingdon, Oxford, 2012)

Eric Jensen, *Different Brains, Different Learners: How to Reach the Hard to Reach* (Corwin, London, 2011)

For parents
To help with children's homework, see Steve Chinn's book below

Marcus du Sautoy, *The Number Mysteries* (Fourth Estate, London, 2011)

Further reading

Rob Eastaway and Jeremy Wyndham, *Why do Buses Come in Threes? The Hidden Mathematics of Everyday Life* (Portico, London, 2005)

Francis Gilbert, *Working the System – How to get the Best State Education for Your Child* (Short Books, London, 2010)

Frank Tapson, *Oxford Study Mathematics Dictionary* (Oxford University Press, Oxford, 2008). This book is designed for students in the 11-16 age bracket, but is useful for parents struggling to help with homework or teachers for whom maths is not their specialist subject. It covers many interesting topics outside the strict limits of the school curriculum.

Carol Vorderman, *Help Your Kids with Maths* (Dorling Kindersley, London, 2013). Very useful and visual manual for parents.

Related disabilities
Roland D Davis, *The Gift of Dyslexia: How some of the brightest people can't read and how they can learn* (Souvenir, London, 2006)

Wynford Dore, *Dyslexia and ADHD – The Miracle Cure* (John Blake, London, 2013). Much criticised by the academic community but quite moving regarding the disabilities of the author's own daughter.

Brock L Eide and Fernette F Eide, *The Dyslexic Advantage: Unlocking the Hidden Potential of the Dyslexic Brain* (Hay, London, 2011)

Naoki Higashida, *The Reason I Jump* (Sceptre, London, 2013)

Valerie Muter and Helen Likierman, *Dyslexia: A parents' guide to dyslexia, dyspraxia and other learning difficulties* (Vermillion, London 2008)

For younger dyscalculics to read
Samantha Abeel, *My Thirteenth Winter* (Scholastic in USA, 2007)

Kathryn Erskine, *The Absolute Value of Mike* (Puffin, London, 2012)

Alexa, Martin, *Girl Wonder* (Hyperion, 2011)

For older/adult dyscalculics
Steve Chinn, *The Fear of Maths: How to Overcome it: Sum Hope*[3] (Souvenir, London, 2011)

General interest
David Acheson, *1089 + All That* (Oxford University Press, Oxford, (2010)

Alex Bellos, *Here's Looking at Euclid: Surprising Excursions Through the Astonishing World of Math* (Free Press, New York, 2010)

Brian Butterworth, *The Mathematical Brain* (Papermac, London, 1999)

Tony Buzan, *Use Your Head: How to Unleash the Power of Your Mind* (BBC, London, 2010)

Ewen Callaway, 'Dyscalculia: Number Games', *Nature*, 9 January 2013

Sarah Cassidy, 'The man who spent millions proving he could "cure" dyslexia', the *Independent*, 31 May 2008

Angela Huth, 'What colour is Wednesday?', *Spectator*, 21 January 2012

Ian Steward, *Professor Stewart's Hoard of Mathematical Treasures* (Profile, London, 2010)

Daniel Tammet, *Thinking in Numbers, How Maths Illuminates our Lives* (Hodder and Stoughton, London, 2012)

Dorothy Wade, 'Professor Brainstorm', *The Sunday Times* magazine, 10 April 2011

Mary Wakefield, 'Intelligence? It's in the genes,' *Spectator*, 27 July 2013

General interest: author's other books
Paul Moorcraft, *Inside the Danger Zones: Travels to Arresting Places* (Biteback, London, 2010)

Paul Moorcraft and Phil Taylor, *Shooting the Messenger: The Politics of War Reporting* (Biteback, London, 2011)

For a further selection in US and UK:
http://www.amazon.com/s/ref=nb_sb_noss_1?url=search-alias%3Dstripbooks&field-keywords=paul+moorcraft

Endnotes

[1] Most research has been done on children. Although figures for adults are thinner on the ground, as is hard evidence of continuation of the condition into adulthood, it is reasonable to accept an extrapolation about adults from research on schoolchildren.

[2] For a useful summary of Butterworth's research, see Ewen Callaway, 'Dyscalculia: Numbers Game', *Nature*, 9 January 2013.

[3] Research scientists in the field may object to my inclusion of the last four paragraphs. These symptoms rarely appear in the research literature and may encourage absent-minded people, for example, to think they may be dyscalculic. I have included them because of my own research and anecdotal evidence.

[4] A former schools inspector and old friend asked me about this section, especially in the context of my life-long obsession with rock singer Del Shannon: 'If there is a correlation between musicality and maths, was Del Shannon a dyscalculic then?' I don't know if he was, but the comment to such a fan was rather hurtful.

[5] See research summary of Dr Marinella Cappelletti: http://www.icn.ucl.ac.uk/Research-Groups/Numeracy-and-Literacy-Group/group-members/MemberDetails.php?Title=Dr&FirstName=Marinella&LastName=Cappelletti

[6] For example, Brock L Eide and Fernette F Eide, *The Dyslexic Advantage: Unlocking the Hidden Potential of the Dyslexic Brain* (Hay, London, 2011).

[7] The only time I did was in the Ministry of Defence. I was discussing a new posting and I suggested something in the field such as in Afghanistan (just before 9/11). I explicitly explained that I should not be posted to the Defence Procurement Agency – that is the

billions spent on buying new weapons – because of my innumeracy. Typically, I was immediately posted to defence procurement.

[8] Living with dyscalculia meant I developed lots of coping strategies, for example hiding my inability to count. Such dissembling helped me in the intelligence world. I learned to break rules by devising lots of ways to enter countries illegally by using different border crossings or simply entering under false guises, a useful part of foreign reporting where journalists are often banned. Dyscalculia also taught me persistence, and a determination to find another way, when conventional routes were impossible.

[9] The obvious question here is what if there isn't such a discrepancy? This may disappoint the many parents and children with 'normal' IQs and less normal (but not disastrous) maths ability.

[10] Fans of the Butterworth approach would say there is very little difference.

[11] As ever, there are scientific papers that argue the converse. When there is no need to calculate time or distance but just to appreciate duration or extensions in space then dyscalculics can perform normally.

[12] I have left the teacher's comments verbatim, but using fingers is of course visualising. Perhaps she should have said 'conceptualising' or 'manipulating'.

[13] Some teachers will argue that use of calculators will undermine the ability to do mental arithmetic.

[14] http://www.mathandmovement.com/activities.html

[15] Some specialists will argue against memory deficiencies and say that the problem is poor teaching. Properly taught (and understood) maths can be better remembered.

[16] Tony Buzan, *Use Your Head* (BBC, 2010) is useful to improve memory techniques.

[17] Buzan's books explain mind-mapping techniques.

[18] Teachers may justly object that this information cannot be expected from pre-school or even nursery children. It will come later in their education. Not knowing such details, at this stage of development, is not necessarily a sign of dyscalculia, but it they do know it, then this is a positive signal that they are not dyscalculic.

[19] Many children do this but it is a possible general indication.

[20] www.suttontrust.com

[21] For more information on the famous chocolatier, see Jessica Salter, 'Word of Mouth', the *Telegaph* , UK, 21 March 2013.

[22] 'Schoolboy inspired to set up his own charity,' *Bicester Advertiser*, 17 March, 2013.

[23] I am referring in particular to the work of the social psychologist Professor James Pennebaker and his 'expressive therapy'.

[24] *Behavioral Genetics* (Worth Publishers, New York, 2012).

[25] A very moving recent example is *The Reason I Jump* by Naoki Higashida (partly translated and also introduced by the novelist David Mitchell). It is written by a 13-year-old Japanese boy to explain how his severe autism influences his life.

Index